RECOLLECTIONS
My life - In and Out of the Military

AN AUTOBIOGRAPHY BY JOHN ACHOR

Cover: Photos, top to bottom
    John Achor standing by PA-18 (Piper J-3 Super Cub)
    North American T-6G (Texan) airborne
        B&W: Row of T-6G planes John Achor flew at Bartow AB, Florida
    Lockheed, T-33A, Shooting Star
        These were the first three planes I flew in pilot training

Color photos by Bob Lesher (submitted for our class reunion)
B&W photos by the author

Unless specifically noted, all images/photos in the book are from the author's collection. To the best of my knowledge, all other photos are in the public domain.

OTHER BOOKS BY JOHN ACHOR

Casey Fremont mysteries

One, Two - Kill a Few

Three, Four - Kill Some More

Five, Six - Deadly Mix

Alex Hilliard thrillers

Assault on the Presidency

Assault on Reason

Memoir

The Iron Pumpkin

RECOLLECTIONS

MY LIFE - IN AND OUT OF THE MILITARY

by JOHN ACHOR
author of THE IRON PUMPKIN

RECOLLECTIONS – MY LIFE — IN AND OUT OF THE MILITARY

© 2023 John Achor

ISBN - 978-1-949601-09-1

All rights reserved. No part of this publication may be reproduced, stored in a retrieval system or transmitted in any form or by any means, electronic, mechanical or otherwise without the prior written permission of the author.

This is a work of nonfiction. Sometimes memories fail and fade, but the author has taken every effort to relate all the events as accurately as possible.

CAVEAT: While many of my military duties were of a classified nature, I have not included anything in this document which has not been published elsewhere in unclassified public venues.

While writing this book, I did my best to inject a chronological sequence to my story, however, when the words flow, I often revert to a stream of consciousness narrative, so hang in there with me.

Author's Links:

BLOG:	https://www.johnachor.wordpress.com/
FACEBOOK:	https://www.facebook.com/jachor1
WEB SITE:	www.johnachor.com
Email the author:	john@johnachor.com
YouTube	www.youtube.com/@John.Achor.Aviating.Author

Library of Congress – Veterans History Project (video narrative)
https://www.loc.gov/item/afc2001001.95861/

1ST Edition, 2023

10 9 8 7 6 5 4 3 2 1

ACACIA IMPRINTS

PRE-PUBLICATION BUZZ

I posted this note on my Facebook page…

I just finished my 200-page autobiography, which I've been working on for more years that I care to count. Now, the real work begins, layout, editing, pagination, converting images for printing, and adding appendices— for pix of planes I flew over 20 years in the USAF, another for short stories that were generated by events in my life and are mentioned in the book and a few more.

Within a couple of days, the post received nearly three dozen "Likes" and here are some of the Comments:

Irma A
    Will be looking forward to reading it!

Deanna D
    I want one.

Karen E
    Congrats on finishing...Count on the sales.

King H
    Count me in for a sale, John. I'm interested in what you have to say. Good Luck.

Robert B
    Let me know when it is published. I will be the first kid on the block to have a copy.

Joyce S
    I hope you'll be including all the good work you did for the ERA over the years. I, for one, appreciate all you and Pat have done!

Elizabeth F
    So pleased with your accomplishment, John. Send a copy along to me at ... with the price tag.

DEDICATION

First and foremost, this is for my wife, Pat. Without her support over the years, I would not be where I am today. Her perseverance was invaluable to our family. That leads to our children, Kathy, Karen and Mark, in whom Pat instilled a love of reading. I credit that fact to their growth and maturity, and I am proud of who they've become and who they are. I am proud to be their father and hope I contributed to their life choices.

Veterans. The first group has to be those seventeen who rode the Rivet Ball off the cliff with me. But I remember you, flying classmates, squadron mates, crew members, mentors and at least some of the staffers. I cannot remember all of your names, and for that I am sorry.

In most cases I was blessed with those I worked for and those who worked for me. These were a talented and dedicated group of officers and non-commissioned officers. Only a very few fell into the opposite category; and in some cases, I have memorialized them in my fiction novels.

Last but not least, I dedicate this book to all of you who have served in our armed forces. I may tend to recall and remember with a male pronoun primarily because there were so few women in the units where I served. And that is most likely our loss.

I wish all of you well, thank you for your service and with the crispest Hand Salute I can muster, I welcome you home.

For those who did not come home, the tear on my cheek is my honor to you. Sleep well, old friends.

ACKNOWLEDGEMENTS

How about I acknowledge every person I have ever known. Each of you contributed to the person I am today and getting me to the = 30 = mark.

Plus, I think I covered it pretty well in the Dedication above.

Thank you, World.

FOREWARD

I published my first memoir, "The Iron Pumpkin," which was limited for the most part to a single instance—the accident that destroyed the RC-135S Rivet Ball aircraft. I referred to "the" because she was a one-of-a-kind airplane.

A reader left a note on Amazon saying the story seemed to be a stream of consciousness. After some thought, I had to agree. But I wonder if many memoirs are written in that style. That's not to say I didn't go back in to patch chinks in the wall or move sections or edit the work. I did all three. I got the major portion of this work written, and over those years filled reams of paper with notes. At the end, I'm going through those pages, typed and written, and scraps of scribbled paper to fill in those chinks.

So, here I am again with a much longer memoir that spans eighty plus years. The major difference is the chronological approach. As I mention above, my hope is that our children's memories will benefit from that approach. They can match ages and locations. That is, if they don't say been there, done that, don't need to read about it again.

I am going to do some of that editing and material insertion I talked about. The new material will be in the [10) February 1957 - July 1957, 8 E. "C" Avenue, Glendale, Arizona] section when I was at Luke Air Force base. I'm not absolutely sure that's when I made these life decisions, but right now it seems to be the logical spot to insert it.

Rather than the usual Table of Contents, I've included the next two sections: First – titles/headings/dates by page number, and second a list of planes I flew while in the USAF.

Following the body of the book, there are three Appendices, A – pictures of all the planes I flew, B – a series of short stories I wrote which are related to the prose in this autobiography, C authors I met, and D – a few personal notes.

The Appendix begins on Page 227

List for locating dates throughout this work (the left column indicates a page number in this book)

PG	PRE-WAR YEARS
1	1) February 1934 – May 1941, 321 Devon Avenue, Indianapolis, Indiana
	THE WAR YEARS
6	2) May 1941 - December 1942, 123 W. 8th Street, Hattiesburg, Mississippi
9	3) December 1942 - March 1943, (street number unknown) Louisiana Avenue, McComb, Mississippi
12	4) March 1943 - August 1943, 604 E. Duke Street, Hugo, Oklahoma
17	5) August 1943 - November 1943, 4964 Kingsley Drive, Indianapolis, Indiana
	POST WAR YEARS
20	6) November 1943 - July 1955, 4931 Winthrop Avenue, Indianapolis, Indiana
32	7) July 1955 - February 1956, 2740 N. Central Avenue, Indianapolis, Indiana
	THE AIR FORCE
	FLYING SCHOOL, FIRST PCS, CONTRACTING OFFICER
35	8) February 1956 - August 1956, 1901 Avenue "H" N.W., Winter Haven, Florida
44	9) September 1956 - February 1957, 304 Johnson, Apt #9, Big Spring, Texas
52	10) February 1957 - July 1957, 8 E. "C" Avenue, Glendale, Arizona
58	11) June 1957 - August 1957, BOQ, Nellis Air Force Base, Nevada
63	12) September 1957 - January 1959, 3208A Oaklawn, Victoria, Texas
	SUPPLY SCHOOL, MULTI-ENGINE
69	13) February 1959 - May 1959, BOQ 6509, Amarillo Air Force Base, Amarillo, Texas
71	14) May 1959 - July 1959, 104 N.W. 12th Street, Homestead, Florida
71	15) July 1959 - February 1960, 2137A Georgia Avenue, Homestead Air Force Base, Florida
74	16) March 1960 - March 1963, 5449C Lemay Avenue, Otis Air Force Base, Massachusetts
	MULTI-ENGINE JETS
85	17) April 1963 - May 1963, BOQ, Castle Air Force Base, Merced, California
88	18) May 1963 - July 1963, (address unknown), Merced, California
90	19) July 1963 - February 1967, 321 Eichenburger Place, Spokane, Washington, Fairchild Air Force Base

		RECON
107	20)	February - March 1967, Castle Air Force Base, Merced Air Force Base, California
118	21)	April 1967- August 1967, 16C Farewell Street, Fairbanks, Alaska
125	22)	September 1967 – November 1969, Eielson Air Force Base, Alaska
		September to October 1967, Visiting Officers' Quarters (VOQ) with family
		October 1967 - December 1967, 5254H Broadway, Eielson Air Force Base, Alaska 98737
		December 1967 - November 1969, 5282 Coman, Eielson Air Force Base, Alaska 98737
		KOREA, SRC
151	23)	November 1969, 2 Carter Terrace, Daytona Beach Shores, Florida 32018
153	24)	December 1969 –December 1970, Det 1, 5AF, Box 799, APO 96570, Osan Air Base, Korea
166	25)	January 1970 – March 1974, Offutt Air Force Base, Nebraska
		January 1970 - April 1971, 305 Chateau Drive, Apt #4, Bellevue, Nebraska 68005
		May 1971 - May 1973, 12273 Turner Circle, Offutt Air Force Base, Nebraska 68123
		May 1973 - April 1974, 707 Sherman Drive, Bellevue, Nebraska
180	26)	March 1974 – January 1976, The Pentagon, Washington, D.C.
		POST MILITARY
		REAL ESTATE
189	27)	February 1, 1976 – March 1981, 707 Sherman Drive, Bellevue, Nebraska
		MORE FRANCHISES
192	28)	April, 1981 - May 1983, 5800 Lumberdale, #113, Houston, Texas 77092
195	29)	May 1983 – May 1985, Dallas, Texas
196	30)	June 1985 - May 1986, 8787 E. Mountain View, Scottsdale, Arizona
		SELF EMPLOYED
196	31)	May 1985 – April 1999 4102 E. Ray Road, #1164, Phoenix, Arizona 85044

	RETIRED, RETIRED
207	32) May 1999 to April 2016, Hot Springs Village Arkansas 71909
	3 Alarcon Way, Hot Springs Village, AR 71909 (rental) May – November 1999
	42 Vega Lane, Hot Springs Village, AR 71909, December 1999 to April 2016
219	33) April 2016 to present - 19301 Seward Plaza, Apt. 218 (later 106), Elkhorn, Nebraska, 68022

Please excuse the chronological date headings. I did this for our children, who traveled with us in the hope it will jog some fond memories of homes and friends they made along the way.

APPENDICES

APPENDIX A   **PAGE 227**

    AIRCRAFT I FLEW

APPENDIX B   **PAGE 234**

    STORIES I WROTE THAT MATCH THE NARRATIVE

APPENDIX C   **PAGE 274**

    PERSONAL, AUTHORS I MET, SPECIAL NOTES

APPENDIX D   **PAGE 280**

    NOTES OF A MORE PERSONAL NATURE

WORD OF WARNING, SOME OF THE BEST IS AT THE LAST ☺

WHAT DID HE FLY, AND WHERE DID HE FLY IT?

Aircraft / builder / description / civilian equivalent (where flown)
* = current in plane at the time, the rest were flown for flight pay

See the Appendix A for pictures of these planes

1. PA-18 Super Cub, Piper, single engine propeller, trainer (Bartow Air Base, Florida) *
2. T-6G, Texan, North American, single engine propeller, trainer (Bartow Air Base, Florida) *
3. T-33A, Shooting Star, Lockheed, single engine jet, trainer (Webb Air Force Base, Texas, training–Foster Air Force Base) *
4. F-84F, Thunderstreak, Republic, single engine jet, trainer (Luke Air Force Base, gunnery training) *
5. F-100A & C, Super Saber, North American, single engine, fighter (Nellis Air Force Base, Nevada, training – Foster Air Force Base, operational) *
6. H-19, Sikorsky helicopter, Chickasaw, staff (Foster Air Force Base)
7. SA-16, Grumman, twin engine propeller, amphibian, staff (Foster Air Force Base)
8. B-47, Stratojet, Boeing, 6-engine, jet bomber (Homestead Air Force Base Florida, staff)
9. KC-97F & G, Boeing, 4-engine propeller, air refueling tanker Boeing Stratocruiser (Randolph Air Force Base, Texas, upgrade training, Otis Air Force Base, operational crew) *
10. KC-135A, Stratotanker, Boeing, 4-engine jet, air refueling tanker, Boeing 707 (Castle Air Force Base, California upgrade training – Fairchild Air Force Base, Washington, operational crew) *
11. KC-135R, Boeing, air refueling training (Castle Air Force Base, California) *
12. B-52, Stratofortress, Boeing, 8-engine jet, bomber (Fairchild Air Force Base, Washington)
13. RC-135D, Rivet Brass, Boeing, 4-engine jet, SIGINT Recce (Eielson Air Force Base, Alaska, operational crew - Kadena Air Base, Okinawa operational crew) *

14. RC-135S, Rivet Ball, Boeing, 4-engine jet, PHOTOINT Recce (Shemya Air Base, Alaska, op crew – Johnson Island / operational crew) *
15. C-47, Skytrain, Douglas, 2-engine propeller, transport, Douglas DC-3, (Osan Air Base, Korea, transport) *
16. T-29, Convair, 2-engine propeller, transport, navigation trainer, Convair 220 (Offutt Air Force Base, Nebraska) *
17. C-131, Convair (T-29 model), 2-engine propeller, transport, navigation trainer, Convair 220 (Offutt Air Force Base, Nebraska) *
18. T-28, Trojan, North American, single engine propeller, trainer, staff (Amarillo Air Force Base, Texas)
19. PA-28 Piper, single engine propeller, (Offutt Air Force Base, Aero Club) *

THE PLANES I FLEW—on page 220, I display a DVD on which I recorded videos of all the planes I flew in the USAF for twenty years. Only my family has (YT) the DVD, but I am considering uploading all those videos to YouTube. If you'd like to view them later, go to my second YouTube channel. Due to YT rules, I may need to load these videos to my primary channel. One big state of flux… I will not delay publication trying to figure out all the YT rules.

My second YT Channel – videos of the planes I flew in the USAF [or, maybe not] .
J Achor aka NOMAD - life in the air and on paper
Bitly (shortened) URL to that channel.
bit.ly/3E7lUKZ

My primary YT channel (writing and flight simulators [and maybe the "I Flew" series]) is here:
www.youtube.com/@John.Achor.Aviating.Author
Note the dots/periods between the last four words in the Handle

= PRE-WAR YEARS
1) February 1934 - May 1941; 321 Devon Avenue, Indianapolis, Indiana

ME AND CHARLIE BROWN

[1] Author's parents

I was born in a log cabin in Illinois and had to chop wood to build a fire so's I could read my law books. No—that wasn't me. That was Abe Lincoln. Sorry about the mistake.

I was however, born at an early age at St. Francis Hospital in Beech Grove, Indiana. Beech Grove is an area on the east side of Indianapolis. My parents are John Raymond Achor and Norma Elizabeth (Long) Achor (picture at left, circa 1927). They named me using a combination of their first names: John Norman. The date was February 27, 1934.

Below, it's me at three weeks of age. My dad was a photographer and part of his duties related to taking pictures of insured buildings.

I remember the house on Devon Avenue, as huge, but then I was small and everything looked big to me. It was a sturdy two-bedroom home, probably build in the 1920s. In those days, home designers paid little attention to refinements. The builders didn't bother to dig out the entire basement to the foundation walls. There was a small area at the bottom of the stairs to provide access to a coal-fired furnace. Dad shoveled out most of the remaining area. That must have been back-breaking work, digging and lugging the dirt up the stairs to the outside.

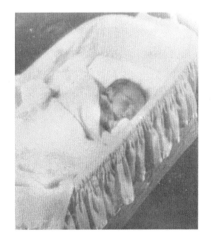

[2] Author – too young to walk

The house sat back quite a distance from Devon Avenue. There was a ditch to cross, sanitary drainage I think – septic tanks emptied or overflowed into it. There was a bridge about three feet wide with railings, and maybe twenty to twenty-five feet long. Once over the bridge, which wasn't always a pleasant trip, there were a number of cement steps to climb to get to the house elevation. Let's digress regarding the bridge. As a young child, parents usually warn against bees etc. by saying, stand still, don't swat them and they will go away.

[3] First home I remember

With these instructions in mind, I bounded across the bridge one day. Lo and behold, a swarm of yellow-jacket wasps swarmed up from under the bridge and surrounded me. I froze and waited for them to leave me alone. Ten or twelve stings later, I determined that these stupid insects were never been exposed to the admonitions of parents. I may not be the sharpest tack in the box, but I ain't stupid, so I took off up the steps as fast as my short legs would allow. In retrospect, I doubt that I had quite that many stings, but who knows...Later, Dad put a gas-soaked rag on the end of a bamboo pole, lit the rag and burned out the wasps' nest.

Back to the top of the cement steps...there was a porch. A left turn and passing through wooden French Doors—pretty fancy, huh? There was the living room. The dining room was to the right, and once into the dining room the kitchen was to the left and at the back of the house. Once in the living room, and proceeding straight ahead, you passed down a hall with a bath to the left and found the bedrooms at the end of the hall. My parents used the one on the left, which was the front of the house, and mine was to the right. Amazing what details remain from the mind of a four or five-year-old.

OFF TO SCHOOL

Warren Township Grade School was a two or three-story brick building located on the far east side of Indianapolis. It was on the southwest corner of the intersection of U.S. 40 – also known as Washington Street, – and Franklin Road. We lived on Devon Avenue, so we were alphabetically (Belmar, Cecil and Devon) three blocks east of the school and two long, long blocks north. Here I encountered the first grade – less than a satisfactory experience.

My parents said all I could talk about the year I started school, was going to school. I don't know if I lacked socialization, if it was separation anxiety or if it was my age. Most likely a combination of all three. Born in February, by September 1940, I was barely five and a half years old. Today, schools probably wouldn't approve such an early start to elementary school. Remember, in the olden days there was no such thing as pre-school or kindergarten.

That made me the youngest in my class all the way through school. Playmates my own age were scarce in the neighborhood. Whatever it was, I took an instant dislike to school. I would ditch school or dawdle so long that one day my first-grade teacher came looking for me in her car.

One day, our marvelous sheriff came to talk to me at school. He said if I continued my truancy, they would toss my parents in jail. Big deal! Didn't faze me. Dad finally adjusted his work hours so he could take me to school, and sit in the back of the room to make sure that I didn't try to sneak out of class. Once I was there for a while, even if I noticed Dad was gone, I would finish the day.

[4] Back yard of our house on Devon

The back yard at home was reasonable large. I remember several crab apple trees. Dad and I would use old golf clubs, one cut down to my size, to "drive" these crab apples out of the yard. There was a goldfish pond there as well, which was another item Dad added to the property. By the way, Mom covered most of the hill out front and leading up to the house with a "rock" garden. There were quite a few rocks in the landscape, but most of the hill had ground cover consisting of various types of ivy.

Access to the garage was by way of an alley. All neighborhoods had alleys in those days. There was a thick hedge along the back property line (although I don't see it in this photo). It made an impression on me, because I can still remember going through the gate in the fence, stepping into the alley and being run down by a bicycle. It was our paper carrier hurrying home after delivering papers. Poor guy, probably in his teens, was scared to death; thought he had killed me. I was more scared than hurt, and he sustained worse injuries than I did. I have no doubt that the bike stopped dead when it hit me, which means he went over the handle bars.

At some point before I was six, I remember Dad taking me sledding in a park. The hill was steeper than anywhere nearby. Since I was so young and didn't take to solo sledding, Dad would lie down on the sled and I got on his back...and off we'd go. Everything was going fine till the last run down the hill. Unplanned, but it turned out to be the last run...we were going full tilt and Dad steered toward what was a bare spot on the hill. The sled stopped dead...we didn't. I was okay after rolling to a stop several feet down the hill. Dad however landed on his face and rubbed his chin and cheeks raw. I remember years later, him telling me he wasn't able to shave for a long time.

WALKING TO SCHOOL IS DANGEROUS

There came a time when I would finally walk to school and stay there on my own. On the way, we crossed Washington Street, which was US Highway 40 on the east side of Indianapolis. It was at least two lanes each direction with a center median. On the way home from school one afternoon, we paused halfway across the highway waiting for the crossing guard to give us the signal to cross the westbound lanes. The guard had those cars stopped and told us to come on across.

The lady in the first car misunderstood the signal and started to go. I was at the front of the column of kids with my friend, Kent Kollman, to my left. We got just in front of the left front fender of the car when she accelerated. I doubt she could have been going five miles per hour when the car bumped us. We went down but the car didn't roll over either of us.

The extent of our injuries was my bloody nose. We were taken to a doctor's office, he pronounced us fit as a fiddle and my mother came to take us home. What a difference in perspectives. Mom was upset, concerned that I was injured. My major concern was the loss of the gloves I was wearing. They were only cloth, but they had simulated "leather" cuffs with "buckskin" fringe. They got bloody from my nose and were thrown away. Alas.

By now, you are no doubt wondering why this section started as: Me and Charlie Brown. Here's the scoop on that story.

HERE'S THE CHARLIE BROWN STORY

I remember having a crush on a little red-headed girl. And that's the extent of the memory... I think her name may have been Judith, but that is the total of my recollections on the subject of early love.

The house on Devon Avenue is the one that spawned my interest in Instant Ralston. Well not so much the cereal itself, but the box tops. That top along with a quarter of a dollar, would bring magnificent premiums from the Tom Mix radio show. For more details, see my short story "The Golden Age of Radio and Box tops" which was published in *Good Old Days* magazine and is published on my web site (www.johnachor.com) and in Appendix B of this document.

THE WINDS OF WAR

My parents decided we needed a phone since Dad would likely be leaving for military duty. I don't think I ever used this one, but it was quite something for folks as far out in the country as we were. It was also an eight-party line, and we were forced to listen to a complicated code of long and short rings to see if it was "our" ring or someone else's.

Several years ago, on the fiftieth anniversary of Pearl Harbor, everyone was asking "Do you remember where you were the day Pearl Harbor was attacked?" At one time, I thought I did, but it turns out I didn't. I remember the pictures in the newspaper that were published a year after the day of infamy. In those days, the government didn't want the populace to know much.

[5] Indiana National Guard Warehouse

Dad served in the Indiana National Guard (38th Infantry Division, the Cyclone Division) for a number of years. His division was activated almost a full year before Pearl Harbor. I remember a cold January night in 1941 when Mother and I went downtown to the Armory to see them off. I saw the big trucks, called six-by's (number of wheels), rolling off into the darkness. It was several months before we joined Dad in Mississippi. Mom knew how to drive and Dad left the car with us—a black Studebaker sedan, circa 1938. (Although I remember it as an armory, it may have been this building. The "pebble" texture is caused by water damage to the 1938 Division yearbook that contains the picture.)

Dad's departure became the subject of a short story I wrote years later. Check Appendix B at the end for *Saying Goodbye*.

= THE WAR YEARS (WW II, BECAUSE THEY FIGURED OUT HOW TO NUMBER THEM BY NOW)

2) May 1941 - December 1942, 123 W. 8th St, Hattiesburg, Mississippi

CAMP FOLLOWERS

Dad's unit, the 38th Infantry Division, Army, Indiana National Guard conducted training at Camp Shelby near Hattiesburg, Mississippi. Mom and I joined Dad there and lived in a private home in Hattiesburg owned by the Andersons. I completed the second grade and part of the third during our stay here. This became the first of three stops as "Camp Followers."

Our first home away from home was a single room in the Anderson house. My best guess about size, again colored by the memory of a young child is that the room was huge...well at least 12 feet by 12 feet or so. I don't remember anything about bathrooms. Obviously, there must have been one, but I don't remember whether there was more than one or not. At least they were inside.

[6] Playmates, Frank Allen and Ann Anderson

The Anderson household consisted of a mother, father, oldest girl (all names long forgotten), a middle girl, Ann and the youngest daughter Nelda. I remember we mostly ignored Nelda because she was younger than we were, "we" being Ann and me. She was my age and we played together quite a bit. I always wondered if Ann got her wish, to have her own horse. Since her parents ruled it out, she begged—perhaps nagged is a better word—her parents for a saddle as a consolation prize. Ann contended she could strap it to the footboard of her tube-framed bedstead and be content for the foreseeable future. She didn't get the saddle either, at least not as long as we were there. This is a picture [6] of Ann and a fellow by the name of Frank Allen. I don't remember him; I imagine he was a classmate of ours.

While living with the Andersons, meals were a complicated affair since we didn't have a kitchen. We took a few meals with the family, but my mother lived in mortal fear I would drink their milk. They served the unpasteurized version. A nearby gas station "served" many a lunch for Mother and me. A six-ounce Coca-Cola in a glass bottle (the original coke bottle shape) and a package of cheese crackers filled with peanut butter was all it took.

For part of our stay, we also ate at a boarding house where food was served family style and good manners were not the rule of the day. I did my best to remember what I was taught. I took a small amount from each bowl as it went around the table and passed it along. When it came time for seconds, the bowls were empty.

[7] Our garage apartment

After several months, my parents located an apartment; actually a "garage" apartment. It was up a flight of outside stairs, above a three or four car garage behind the main house that belonged to the Strahans. I don't recall their first names, but Mrs. Strahan always called her husband, "Husband." Maybe he really didn't have a first name – or maybe that *was* his first name. The garage was behind and to the west of the house.

LIFE IN THE SOUTH

The apartment consisted of two bedrooms as well as a living room, eat-in kitchen, bath and a screened porch both front and back. Two stairways led from ground level, one to the front door opening into the living room, and the other to the back porch, which led to the kitchen door. It was also only a couple of blocks from my grade school.

We kept rolled up newspapers and a "Flit" gun near the back door. Flit guns were one of those old garden tools used to spray insecticide on your vegetables. It had a tubular barrel with a T-shaped pump handle extending from the tube. Toward the front end and below the tube there was a reservoir like a tin can that held the poison. The noxious liquid was for large insects, cockroaches to be specific. During the hours of darkness, especially when we were gone, they "owned" the house. On coming home and reaching the back door, we would arm ourselves with the rolled newspapers and poison dispenser, unlock the door, reach inside to flick on the lights and leap into the kitchen to do battle with these giants. Okay, so they weren't the size of the ant in the sci-fi movie "Them," but they were large enough we could hear them scrabbling over the wallpaper at night when the lights were out…and they were plentiful.

Mr. Strahan maintained a garden on the west side of the garage (the garden area was behind the person taking the picture above of the garage, see image 7). It must have been a couple of acres, and the only crop I remember him planting was turnips. Apparently, no one in Mississippi liked turnips, but they did enjoy the greens. So, Mr. Strahan kept the leafy portion and threw the turnips away. That was okay with those of us who conducted dirt clod fights. We protected ourselves from incoming missiles with garbage can lids.

And when there was no clod at hand, we didn't mind grabbing a turnip or two by the exposed leaves as a suitable substitute. In retrospect, I doubt that the greens were salvageable after jerking the turnip from the ground to hurl it at the "enemy."

I can't recall the names of the folks across the street to the north of us. It was a large old southern home. There must have been a male about my age, as I do recall playing with someone from over there. The property was large enough that they raised or at least kept quite a number of dogs. Don't recall how many, but there were several dog houses in the kennel area behind the house.

I'll have to estimate the number and it will probably be too high. Thinking back, things and numbers always seem larger than life to small kids. I'll guess there was room for eight to ten dogs. The kennel space came in handy when hiding from the younger sister of the boy I played with. In an effort to ditch her one day, we raced back to the kennel area and ducked into one of the dog houses. At that age, I knew nothing of dogs. I didn't even know that fleas existed. Needless to say, I learned what fleas were and how their bites could itch. I may have told my mother, or she may have noticed the little critters on me. They were dispatched via sheep dip or a good Saturday night bath, I don't remember which.

At some point, which may have been years later I learned that we lived a little over a block from the Hercules Powder Plant. It was a huge factory producing munitions for the war effort. Good thing it didn't blow up while we were there—could have ruined our whole day.

3) December 1942 – circa. March 1943, Louisiana Avenue, McComb, Mississippi

VAGABONDS MOVE ON

Dad was still assigned to the 38th Infantry Division, but portions were moved to another location. He was stationed at Camp Van Dorn which was near McComb, Mississippi, and I finished the third grade and spent part of the fourth in McComb. I've searched every record I have and could never locate the street number for this address.

Our home was another typical old southern style. One floor with a hall running from the front door to the back door—shotgun style. It was what I think of as a "double," but it had a hallway instead of a wall separating the two parts of the house. The owners lived in the north half. The street was oriented northwest/southeast.

We had the south half of the house. I remember a living room, a dining room and kitchen all in a row, front to back. Locations of the bathroom and bedrooms escape me.

We were only three or four houses northwest of a main drag, Broadway, which ran north into the center of the city. I have no recollection of the downtown area. My school was six to eight blocks along Louisiana Avenue, northwest of our location. I can remember walking to get there, but nothing else about the school.

The most memorable recollection of this house was Christmas time. We put a tree up in the living room but didn't have much in the way of ornaments. Mom made popcorn and using needles, we strung it on thread and placed the "garlands" on the tree like strands of tinsel. We went out for the day, and on our return sometime after dark, we discovered that the popcorn turned from white to black. A closer inspection revealed that not only did the color change...but also the popcorn strands seemed to be alive with movement. They were. The popcorn was covered with ants determined to carry off this once in a lifetime treat. From the same invasion, we learned ants also enjoyed nylon. They invaded my mother's dresser to get to her stockings.

I may be mixed up on dates, but about this time Dad bought a new car. Trading in the Studebaker, he purchased a new 1942 blue Plymouth two-door coupe. This automobile was one of a small number of autos produced this year. Due to the war effort, civilian car production ceased and did not resume until 1946; this body style carried over. We kept it long enough I learned to drive in the old Plymouth.

VACATION TIME BETWEEN CAMPS

[8] The author taking a picture of the picture taker – the '42 Plymouth in the background

We drove back to Indianapolis for a short vacation before Dad reported to his next assignment. We traveled along the Mississippi coast before heading north. This is a picture of me taking a picture of my Dad taking a picture of me. Don't know what happened to the ones I took with my Kodak Brownie camera. We were in Biloxi, Mississippi.

[9] An oyster boat, the Tony Jo

While there, we also visited the shore where we got a ride in an oyster boat, the Tony Jo.

[9a] The author at the tiller

While "at sea" (actually the Gulf of Mexico), the skipper put me to work. There I am struggling with the tiller.

Back on the road, we got as far as Jackson, Mississippi in the south-central part of the state before stopping for a major purchase. The new Plymouth was delivered without a heater. We were freezing and Dad stopped at a car dealer and asked the dealership to install one.

REMINDERS OF HOME

These images were in a foldable "leatherette" holder. Dad carried this reminder of home with him throughout the war.

The backs of the pictures were inscribed with the age and year of the photo. Mine read "10 years old, March 1944" and Mom's read "39 yrs old, Mar. 1944."

[9b] Photos of the author and his mother

My memory is not that good. The photo case is sitting on a cabinet about four feet from where I am typing this entry. It's in darn good shape considering the folder is at least 75 years old.

4) March 1943 - August 1943, 604 E. Duke Street, Hugo, Oklahoma

STRANGE DIGS

I finished the fourth grade in Hugo, Oklahoma. Dad was stationed at Camp Maxey, near Paris, Texas, which was south and just across the state line from Hugo. A new division for Dad (99th Infantry Division – Checkerboard Division), more Army training and a new town for us.

This house had the strangest floor plan I've ever seen. My powers of description fall short, so I'll include a sketch. Four to five steps up from the living room to a platform and through the door gave access to the stairs to the second floor, or through another door, to the bathroom. If you opened the door from the platform and rushed in too fast toward the bathroom, you were likely to go head first into the tub. No railing there or in the living room.

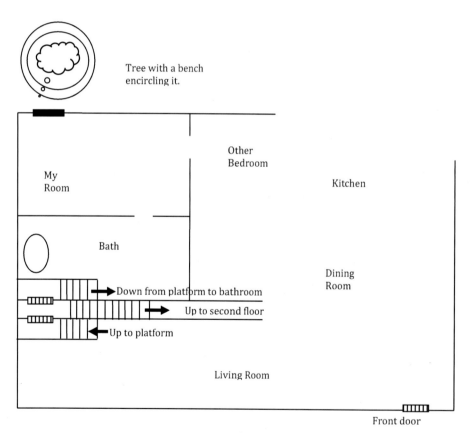

Not to scale and room locations approximate
[9c] Hugo, Oklahoma floor plan

I've always wondered how a designer came up with such a weird design for a house. In retrospect, I can guess that the second floor was added after the building was complete. The stairs in the middle of the house were the only way to provide access to the upper level. For the life of me, I cannot remember anything about the second floor.

My bedroom was at the back left—and it would be the southeast corner of the house. Seems to me the street we lived on was oriented east-west, and the house was on the south side of the street. The neatest feature of my room was the windows. They opened onto the back yard and provided an easy route to the outside. I think I could drop from the window ledge to the ground, but that's another fact I'm unable to verify in my mind.

However, I made it, I had easy access to the back yard where a boy from the house next door, Ricky Roundtree, and I played.

DOWNTOWN IS EVEN STRANGER

Walking to the center of town was a easy jaunt, several blocks to the west. Hugo was a typical "old town" of the era. The main area of "downtown" was the town square. The only specific feature I recall was the Trading Post—a non-descript old timey general store that bought, sold or traded whatever you can imagine. I particularly remember a very large Mason-type jar in the front window. If was filled with sets of false teeth.

If one were in need of dentures, all one had to do was stroll into the Trading Post, select a set from the jar and try them on. If they didn't fit, potential buyers were welcome to drop the first set back into the jar and select another set of teeth. Gross to the max.

Time for Dad's division, the 99th Infantry (Checkerboard) Division, to enter the war approached. As they prepared to head to England and the European theater, Mom and I packed up and moved back to Indianapolis. She drove the car from south Oklahoma to Indiana. I'm not positive of the route, but I remember this vision as having occurred in Kansas. As we hummed along the highway—remember there were no interstates in those days—I can picture water along the road. Kansas must have had an overabundance of rain because I was looking at flooded fields as far as I could see on both sides of the road.

I never thought of Mom as a particularly strong individual until I recall incidents such as this. Here she was with a nine-year-old in a car virtually surrounded by water and no doubt, thinking about a husband who was packing his belongings into duffle bags and preparing to go overseas where the fighting was.

ANOTHER DIGRESSION

[10] First Grade picture of the author

Before I pick up my story, I'm going to digress, rather regress back a few years and add some photos. These three are from the first grade in Indianapolis. It was getting close to the time we were to leave and become camp followers when I took my Kodak Brownie to school with me. That camera used 127mm film and was a basic point and shoot unit. In this one, a classmate clicked the image of me, which was a bit askew. Note the cool dude in long pants and suspenders.

[11] Miss Wampner, First Grade teacher

I couldn't get away without including a snapshot of my first-grade teacher, Miss Wampner. As far as I can remember, she was a good teacher and went out of her way to help me get used to school. As noted earlier, that was a job and a half. I remember a day when halfway to school, I was having serious doubts about continuing the trek. Many blocks down the street I saw a car and decided I would wait for the car to pass before crossing the street. Bad timing—it was Miss Wampner and she picked me up and took me to school.

Finally, I'll include a group shot of classmates. I have no idea who any of them are.

But, maybe, just maybe the little red-headed girl is in this picture …

Maybe …

[12] Author's First Grade classmates

These shots are from McComb, Mississippi about the time we were leaving for the last time. I took this one of Dad and Mom. I was probably using his camera.

[13] John R. and Norma E. Achor

Mom took this snap of Dad and me.

[14] John R. and John N. Achor

I don't know where this fits, so this is just as good a spot as any. I know there was a death notice from an old newspaper. Whether I found it or it was shown to me as part of a discussion, I learned that I should have had an older brother. He was stillborn about a year before I came along. I think he was never named. Over the years, I've often thought about him and wondered what it would be like to have a brother. I'll never know, will I?

That's enough of nostalgia within nostalgia. Let's get back to the rest of the story.

= POST WAR AND THE REST OF THE GRADE SCHOOL YEARS, AND SO, ON TO HIGH SCHOOL, COLLEGE AND MARRIAGE

5) August 1943 - November 1943, 4964 Kingsley Drive, Indianapolis, Indiana

BACK TO THAT FIRST HOME I REMEMBER

[15] 1930s furniture on Devon Avenue

When we left Indianapolis, my parents rented our home on Devon Avenue to a couple. We left all our furniture behind for the renters and boy did they use it … and use it. The couch had a cloth throw covering the back and the cushions. When we returned, the throw was still there, but it was worn so thin it was in danger of disintegrating. This is the picture of the only furniture I remember my folks bringing along. It is a 1928 RCA Victor Victrola. This was a spring-driven mechanical device, no electricity required. A crank handle on the right side wound a spring to provide power to the turntable. The Victrola played 78 rpm records. The arm was heavy and the needle was a half-inch long and made of steel. A fully wound spring would play two or three records before it ran out of power. After the war, Dad tore the guts out and replaced them with a modern electric turntable. So much for antique furniture.

Back in Indianapolis, we moved in with my aunt, mother's sister; my aunt's two children, Mom's mother. In those days, itinerant photographers roamed neighborhoods offering and to take pictures of kids on the photographer's pony.

[16] Author on his trusty steed

I wish I'd asked questions of my grandmother during that time. I know so little about her life and her struggles. Both my dad's parents were dead by then and I thought Grandma was my last remaining grandparent.

Little did I know my granddad, Mom's father, was still alive. I discovered this on the day of his funeral. Mom, her sister and her mother were "dressed for church" when mother announced, "We are going to your grandfather's funeral."

They went, and we, my two cousins and I, stayed home. Later, I learned grandpa was a drunk. Blunt, and difficult to sugar coat. He almost never made it home on Friday night with his paycheck—too many bars along his route.

Grandma kicked him out of the house. Some of the questions I could have asked: how long ago did she throw him out? How did she survive with two children to support? Did Mom and her sister pitch in? Did Grandma earn money? How? Did he lend a hand? I didn't ask. Now, my grandmother, and my mother are both gone and I will never have those answers.

[17] Author on his first bicycle

During this period, I attended Public School #91, which was a short walk from my aunt's house.

Here, I completed the fifth grade, learned to ride a two-wheeler bicycle—my first was a 15-inch wheel, used bike that had a terrible repaint job; someone used a mop to slather on a coat of blue.

My male cousin and I participated in rock fights with other neighborhood kids. Fortunately, teams maintained enough distance between themselves and the "enemies" there was little danger of being struck by flying missiles.

This picture was taken in the driveway on Kingsley Drive. I'm front left, cousins Marjorie and Paul are next to me. My Grandmother Long (Mom's mother) is behind us. That's' the famous 1942 Plymouth Mom and Dad owned behind us.

[18] Author (front left) with cousins and grandmother – in front of '42 Plymouth

Pat, my wife, lived two blocks south and one block east of this home. She also attended P.S. 91, and we never met; more later.

There were a couple of other memories from this era. I have no idea why someone decided to buy a live chicken. You can't eat chicken until it's dead, so my grandmother inherited the task. Rather than wringing the hen's neck, she got a hatchet, put the bird's neck on a stump in the back yard and whacked the head off. When she dropped the body, I got an up-close of the old saying, running around like a chicken with its head chopped off. That bird made two laps around the yard before it dropped dead.

The other item I remembered was in that same backyard. That cherry tree must have been twenty feet tall the year they decided to remove it. We picked the cherries from the low hanging branches and then cut the tree down. After harvesting the rest of the cherries, it was cut into small enough pieces to be dragged away.

The next project was to pit the cherries. Using old wire hair pins, we pitted and pitted and pitted for what seemed like an eternity. We did at last finish and enjoyed any number of fantastic cherry pies.

6) November 1943 - July 1955, 4931 Winthrop Avenue, Indianapolis, Indiana

A NEW HOME

Anticipating the end of the war in Europe, my mother went on a expedition to locate a house to rent. In fact, we may have already gotten a letter from Dad indicating the entire division would be rotated back to the States. That was true in general, but Dad was tapped to remain and process all the last-minute paperwork required. The division was gone, and Dad was working in the Corps headquarters. Again, he was selected for the "honor" of being the last man out. He even had the opportunity to sign documents as the Corps Commander. A typical Corps consisted of tens of thousands of troops. I think Dad said by the time he took over, they were down to a couple of dozen.

Mom found a house a few blocks from my aunt's home. Having placed a deposit on the rental and signing a lease, she and I went there to clean it up before a move-in. Well, she was cleaning and I was roaming around the neighborhood and the yard—a real yard surrounding a detached house. Satisfied with what I saw, I wandered into the garage, which seemed to be a trash dump for the previous tenants. When I say trash dump, I'm being generous. It was more like a land fill. I went into the house and notified Mom there was another chore awaiting her in the garage.

I did my best to describe the junk out there, but she thought I was exaggerating. At last, she followed me out to the garage and the enormity of the task ahead struck her, she uttered something. I don't remember her words; she was not one for cursing, but this time I'm sure not much of what she said was repeatable.

The leasing agent was adamant when Mom called and complained—lease was iron-clad, there were no refunds on deposits and basically conveyed an attitude of, "that's tough." I don't know what she said, but the agent cancelled the lease, gave her the deposit back and we went back to house hunting.

The next home she located was a little over a mile due west of my aunt's house. There was a string of doubles—two single family dwellings, side by side in the same "house." These were one-bedroom affairs with a living room in front, a small dinette area and a dining-car kitchen beyond that. At the end of the kitchen, the only bath contained a tub—showers were unheard of in those days, at least in our neighborhood. Past that, and at the rear of the house was the bedroom. A door in the dining area, led to the back door and stairs to a basement that was divided into three areas. In the larger part which was under the front two-thirds of the house, was the furnace (originally coal fired and later converted to oil), wash tubs, room for a washing machine, some storage area and a door to the coal bin. A wall paralleling the stairway had a door that led to the room at the back of the basement. Initially, I slept on a roll-away bed in the living room. Later, this basement room became my bedroom.

We were on the north side of the northern most building of the six or seven of these doubles. Our side yard abutted a gravel street leading past our unit and bending to the right (south), providing access to the row of garages for the homes. Nothing fancy here, all single car garages, plain interiors and these garages formed a single unit that was a part of the wall separating us from a plumbing business east of us. The man or family who owned the plumbing business was also our landlord. The business yard was quite large and contained a railroad spur line inside the area. It was easy to skinny under the large wood gates that closed over the tracks.

Just east of the land used for the business, was a single-track railroad running north-south—The Monon Line. Like most noises, we became used to the steam engines rolling south to Indianapolis stations and north to the city of Monon and beyond. The trains gave us the opportunity to place a penny on the rail and wait for a train to pass. The steel wheels flattened the coins to about double size and often it was hard to locate them after the train cleared the area.

Also, my first love as a teen moved into the other half of the double (two homes in the same building) we occupied. It was a short-lived situation because she shortly began another trek to follow her dad since he was still on active duty. Her family moved, and we lost track of one another. Oh, woe!

FDR (ANOTHER ASIDE)

There was a large open area on our side of the train tracks. We converted it into a baseball area. I remember being out there playing when the death of President Franklin D. Roosevelt reached us. I was saddened by the news; Roosevelt was the only president I'd known.

Years later, the train came in handy for me. We were staying at an uncle's fishing cabin somewhere north of Indianapolis and for some reason I "needed" to get home before our scheduled departure. I think I was getting bored. I suggested catching the train in a small town nearby since it was the one that ran by our house. I could get off at the 38th Street, station and walk home. Since we lived at 50th Street, it was about a mile and a half walk. The train ride and the walk were simple, convincing my mother I could do it on my own was nearly an insurmountable problem. I convinced her, rode the train, walked home and all was well. I have no idea what my "pressing" need was other than to do what I did.

WW II was still in full swing and it was the day of V-mail, short for victory mail. No electronics, this was a process that reduced the size of a piece of paper to about four inches by five inches—sometimes hard to read depending on the handwriting—and folded on all four sides to form its own envelop. No extra space or weight. As important as mail is to the soldier in the field, cargos like munitions took precedence.

I have a mental picture of myself, sitting at an orange crate—an obviously suitable substitute for a desk—located in the living room closet. I would drag a dining chair in there and compose letters to my father. I'm not sure they made any sense and were usually thank-you's to him for the war memorabilia he sent home.

THE END OF THE WAR

Sometime during these years, and after the war I remember an incident about my father and me. We were out somewhere, doing something and I have no idea what. I only remember one short part of the day. First, you need to understand something about comic books of that era. There were two types. The most popular one was "bound" using staples like many magazines. Its popularity was the price—it cost only ten cents, one thin dime. The other style of comic book was put together without staples with a binding similar to books today that are referred to as "perfect bound." That is, they had a spine and the pages were glued to it. These were the coveted ones, costing the princely sum of fifteen cents. Seldom did I deign to ask for the more expensive versions.

On this day, Dad stopped at a corner drugstore and took me in. He treated me to a chocolate sundae—vanilla ice cream, chocolate syrup served in a fluted glass. I would be willing to bet we sat at a round table on drugstore chairs. These consisted of round wooden seats, backs formed by heavy gauge wire formed into loops and legs made of several pieces of the same wire twisted into braids for strength.

"Would you like a comic book?" Dad asked. Nodding vigorously, I headed for the comic book rack—a round affair with several levels to hold the magazines and mounted on a swivel base. No need to move around the rack, simply remain still and rotate the stand until the next batch came into view. I selected a comic book, one of the expensive variety and approached our table.

I held the comic up so Dad could see it and he said it would be fine. What more could a kid ask for—a trip to somewhere, to do something and being treated to a chocolate sundae and a fifteen-center.

I finished the sixth through the eighth grade while living on Winthrop and attending Public School #80. There were two teachers here with the same last name, which I can't recall. One was our math teacher and the other was the principle. Turns out, our principle taught school way back and Dad had her as one of his teachers.

Public school, P.S. 80; nearly a mile walk was the sixth-grade school I attended. This was also the scene of my short-lived sports career. At recess we played softball occasionally. I was no good at this sport, but I finally got a hit one day. I was so excited I slung the bat as I took off for first base. Unfortunately, the bat struck a classmate in the head. The injury wasn't serious, but I gave up softball.

Indoors, we tried gymnastics. One of the exercises was a diving forward roll over members of the class who were down on all fours. The class jock could easily clear five to six classmates without breaking a sweat. I barely got over two and landed on the top of my head. I forgot to tuck my head so I would land on my shoulders. End of my gymnastics stint.

Also, indoors, the gym teacher would divide us up into two teams for the infamous dodge ball game. The only day I remember was the day the class jock caught me square in the face with a ball he heaved with all his might. I saw stars for the remainder of the class period and that was the end of my days as a target for dodge ball—I avoided it like the plague. Years later in a high school reunion newsletter, I saw that the old class jock died. I hadn't thought of him for decades. I wish I could have had positive thoughts about him, but I didn't.

I graduated from Broad Ripple High School which was on 63rd Street, on the north side of Indianapolis. In those days, that was the north edge of town. Today, it's probably nowhere near the city limits. I walked three blocks to catch a streetcar which ran on railroad type tracks and was powered by electricity from overhead wires. I boarded the streetcar at 50th Street on College Avenue, and it was a mile and a half ride to high school.

FIRST JOB

During my sophomore or junior year, I had a paper route. Figured I could make my fortune and retire. I serviced about fifty customers and delivered the evening paper, The Indianapolis News. Later it was integrated into the Indianapolis Star, the morning paper. My three-block route was only a block west of our street, Winthrop Avenue, and they dropped the papers at the corner next to our house. Neighbors complained about the mess we left behind—wasn't me, must have been the other carriers—and the newspaper distributors relocated the drop-off spot. It was a good mile east of us on 49th Street. It was quite a hump on a bicycle. The worst part was that 49th Street was a two-lane road with no shoulders. Keep this pick-up spot in mind; it will play a role in another part of the story below.

Saturday morning was paper route collection day. Talk about cheap, six papers a week and the cost was twenty-five cents. This activity gave me the chance to see my customers face-to-face. I only remember a couple of the faces—at one house there was a very attractive wife and daughter. She's the one I called, the daughter that is, and she wasn't at home, which broke me of trying for a date. The other house I recall was one where an attractive young wife always came to the door in a white T-shirt—a major thrill for a teenage boy. There was another face-to-face encounter that cost me money. I rolled papers so they became a self-contained delivery vehicle and tossed them from the sidewalk up to the front porch. This particular house had a storm door with two glass panels around two feet square each. I lofted the paper toward their porch and as soon as it left my hand, I knew the spot it would hit. You guessed it, the exact center of the lower glass panel. I started up the walk to the house and the owner opened his front door. In fell the paper, and the broken glass. I said if he'd get the glass replaced, I would pay for it. That took the profits out of several weeks of paper deliveries.

My second job was also in the news business, sort of. I worked for the paper I had delivered, but in an office in downtown Indianapolis. As the office gopher, I chased around downtown area stopping at department stores and anyone else who was advertising in the paper. I picked up ad copy, returned it to our office where the artists converted the copy into printable material. Not enough newsprint in my veins to sustain me, so I looked for other opportunities in college.

ALL THE CONVENIENCES OF HOME

These were also the days of the ice deliveries from a twelve-foot stake bed truck. Our home had an ice box instead of a refrigerator. We were furnished a foot-square card with a different number on each of the four edges. By placing the correct number—25, 50, 75 or 100, toward the top and putting it in the front window, the iceman could haul the correct amount to the door without asking. I can still see him, a piece of burlap on his shoulder to protect him from the cold and wet ice, hefting a block of ice with his ice-tongs to his shoulder and coming up the sidewalk. Sometimes, we would hide out in the yard near the truck and when he entered a house, we would snitch small chunks of ice that had broken off larger pieces. He would "chase" us away, but I don't think he was all that serious—my guess is that he did the same when he was a kid.

What happens to the ice in the ice box you say? Well, it melts. The water runs from the top icebox space, a separate door, down a hose to a catch pan under the fridge. If you left it too long, you got a puddle on the kitchen floor, so it had to be emptied on a regular basis. No ice cubes in those days. If you wanted ice in a drink, you opened that top door where the block of ice resided and went to work with an ice pick. Amazing how an experienced chipper can extricate just the right size piece of ice for a glass. Rather like a sculptor working in granite.

THE BEGINNING OF MY MILITARY CAREER

In high school, I enrolled in the Army ROTC program and found the Army sergeant's name was almost the same as mine; spelled it differently – Aker. He always ragged on me over our names. I gave it back in return. There was a 1,000-inch range in the basement below our classroom. The Army shipped a dozen or more 22 cal. rifles—packed in Cosmoline, a rust preventive. I was chosen to participate in the cleaning process; getting that gunky, greasy mess off the weapons was a messy challenge.

I earned a marksman badge with those rifles and attained the rank of Cadet Captain (three round disks in a row) and served as a platoon leader.

In my senior year I began smoking. After school, three of us—Ed, Ken (last names redacted to protect the guilty) and I would go down a small hill to the edge of White River. We crossed the street in front of the high school, went over the streetcar tracks and disappeared down toward the river bank. The river ran along the north side of 63rd Street. We would puff our way through a couple of cigarettes each, then climb the hill back to the streetcar stop. We deluded ourselves, believing our parents would never smell the odor of tobacco on us. It was decades before I finally kicked that habit and realized that smoke permeates everything and follows a smoker where ever he goes.

I was shy during this time. Believe it or not, I was the one who called a girl for a date and hoped she wasn't home. One go-around at that—and finding she was not at home—was enough to cure me of calling young ladies. That was the young girl on my paper route. During registration for my senior year, I found myself in line behind an attractive girl, one I had a crush on. Locker assignments were made during this process, and I decided that no matter how long it took in line, I would stay there—forever if need be. The thought of having a locker next to Carol for an entire school year was too much to pass up. I stayed, we got lockers next to one another and I saw her a couple of times every day for those semesters. All that and I never got the nerve to ask her for a date.

Attending a school evening event, I was with a guy I knew and we ran into Carol and a friend of hers. She invited both of us to her house after the event. We "hung out" and played ping-pong—and left. I guess I never got the hint, because I still didn't ask her out and "worshipped" her from afar.

My dad remained a member of the 38th Infantry Division (National Guard) during these years. Each summer, the division went on two-week maneuvers in

Michigan. One year, he convinced be to go along as an Orderly—a fancy term for someone who would make beds, shine shoes, and be a go-fer for the officers in one of the barracks. It included three slops (meals) and a flop (cot) for me and they paid be a grand total of $15 total, each and every week. Last of the big spenders. Considering the number of officers in a single barracks, it must have cost each of them about a buck a piece to hire me.

[19] Author's first car; 1940 Ford Club Coupe

Figuring I would need my own transportation to college, my folks bought me my first car, a 1940 Ford Club Coupe.

I started college in June 1951 having graduated from high school the same month. Butler University is in Indianapolis and I enrolled in Air Force ROTC (Reserve Officers Training Corp), which was the only service unit available. It also was accepting only pilot training candidates. That gave me a 1D Selective Service rating which would keep me out of the military until graduation. No sense going to Korea and getting shot at any sooner than necessary.

AND ON TO COLLEGE

Butler University was an in-town school and catered to local business people. Many of my accounting classes were scheduled at night. The campus was located around 49th and 46th streets on the west side of Indianapolis. Many of the buildings are still there and the campus has grown exponentially.

I joined Kappa Sigma fraternity, and I discovered it had a racist charter, limiting membership to white only. West coast schools and Hawaiian chapters wanted to eliminate the restriction so they could pledge Asians and Hawaiians. I remember speaking at a chapter meeting in favor of dropping the "whites only" clause and the national office dropped the clause shortly thereafter. I doubt that action was due to me alone, but enough chapters supported the position and forced the national organization to capitulate.

I think it was in my senior year at Butler University that Dad invited me to join the Masonic Lodge. He had recently been inducted and was a 3rd Degree, Master Mason and he later went on to become a 32nd degree Scottish Rite. I joined Masonic Lodge, Oriental #500 (ca. 1954); and maintained active membership for many years even though I was not living in Indianapolis.

Years later, I let my membership lapse in the late 1970's, due to a lack of attendance, interest, and awareness that the Masonic Lodge was racist and extremely slow to adapt to the changing times.

I met my wife, Pat, while I was in college. Here's a strange batch of coincidences that put us in touch with one another. Pat, born in St. Johnsbury Vermont, grew up in Buffalo, New York until she was nine. Hitler was running rampant in Europe and our country decided it needed an atomic bomb. Pat's dad was a chemical engineer and was transferred to Linde Air, a division of Union Carbide in Indianapolis where they lived in three different locations. He became part of the Manhattan Project. Due to that same war, Mom and I followed Dad all over the South in the early 1940s because his National Guard division was activated. We finally returned to Indianapolis.

[19a] Author with his fiancé, Patricia Hendry

Our paths crisscrossed without meeting—Pat and I attended the same grade school when I was in the fifth grade and she was one year behind me—she was younger, they didn't hold her back. I picked up newspapers for my delivery route across the street from her house. I could see it and could have hit it with rock from that pick up point. We went to the same high school, and didn't know one another during any of that time. Finally, we met while we worked at American United Life Insurance Company (AULIC). I worked there part-time while I was in college, since this was my dad's employer— the same one that let Dad arrive late because he'd been sitting in the back of my first-grade room. Pat was a full-time employee there and we finally met in person.

[20] Second automobile, 1946 Mercury Coupe

Somewhere in this time frame, I got my second automobile. This one was a 1946 Mercury Coupe.

While working at AUL doing odd jobs, another of the work-gang gave me a "ride" in a wheeled canvass trash cart. The guy rolled me into the president's office—the president was in his office and Dad heard about my wild ride first hand.

This is my senior picture [21] from the Butler University year book.

A week in the middle of June, 1955 was a busy one. I received my B.S. degree from Butler and was commissioned a 2nd Lieutenant in the USAF; Pat and I were married in that same week. That was 67 years ago (as of 2022).

Christmas parties at AUL were notorious. The day began with wing parties. Since there were two wings to each of the three floors, floaters like the gang I worked with had a wide range of choices. Later, everyone went to the third-floor auditorium for dinner. One guy I knew, missed dinner three years in a row—too inebriated to make the finale.

[21] College Senior Yearbook Photo

One of our favorite pastimes as teens, was to cruise the drive-in hangouts; the restaurant type that is. There were at least four on the north side of Indianapolis and another less frequented one out west in Speedway. That was a suburb near the Indianapolis 500 Motor Speedway—the home of the "greatest spectacle in racing" every Memorial Day.

Remember the old photo booth? Do they still exist? This is a shot of Pat and me hamming it up, when we were dating, probably 1954,

I was earning next to nothing as I finished college. I managed to put together enough cash to purchase a wedding ring set. I planned a special evening at the Indiana Roof, a dance spot in downtown Indianapolis hoping it was romantic enough to get a "yes" to the question I was about to ask.

[21a] Old Photo Booth

This is the cover of a souvenir album prepared by the photographer at the Roof. The stars you may see in the pic are not the real ones, they're painted on the ceiling since the place was used year around.

The Album cover lifted up revealing a photo of the two young lovers.

[21b] Indiana Roof

And here they are, enjoying a 7-UP and other soft drinks. Remember, this is Indiana and we were both under the drinking age. Shucks!

There I sit, cool dude in his flat top crew cut, unable to believe that in a short time I am going to ask this drop-dead gorgeous young lady to marry me.

With the ring box in my pocket, we enjoyed the evening and I held off on the proposal until we got to Pat's home.

[21c] Author and Pat, soon to be engaged

Isn't a 1948 Mercury Club Coupe the most romantic place you can imagine for a proposal?

It must of have been because Pat said yes. The beginning of a 67-year plus odyssey, which is still going as I write this.

You might notice specks on my suitcoat. I not a sloppy drinker, those dots are over much of the photo. Old age spots I imagine. Speaking of the suit, that jacket was a one-button rolled lapel. A single button which fastened more like a French cuff than a regular coat. The lapels rolled rather than being pressed flat. Way cool, back in the day.

Newspaper clips from that era. At the left is the announcement of the Marriage License we applied for.

On the right is the engagement announcement from the Indianapolis newspaper

[21e] Engagement announcement

[21d] Notice of marriage license

I did my best, but newspaper print doesn't age well and these are as old as our marriage.

This Indianapolis Star paper's announcement of our vows, was laminated and mailed in the blind. I think we did pay the requested amount.

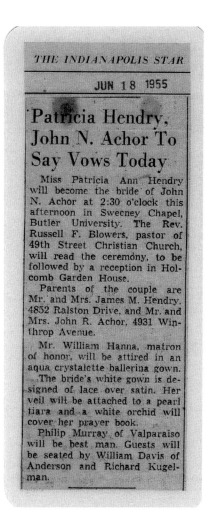

[21f] Wedding announcement

Here we are leaving Sweeney Chapel on the Butler University Campus in Indianapolis.

[21g] Happy couple leaving Sweeney Chapel

When we arrived in Daytona Beach, Florida, we found that Pat's grandparents put a note in the Daytona Beach Journal, dated June 24, 1955 celebrating our honeymoon in Daytona.

> ... Pat and John Achor are here from Indianapolis for several weeks' stay with Pat's grandparents, Laura and George Palmer on Ridgewood Ave. This isn't an ordinary visit. They were married last Saturday, right after John was graduated from Butler College, and this is their honeymoon. Pat has been coming to Daytona for seven Summers now.

Later, look where Pat decides to live while I was on an unaccompanied overseas tour.

[21h] Daytona Beach paper notice

I loved Laura and George Palmer like they were my own grandparents. I came up a bit short in the grandparent department. My maternal grandmother I mentioned earlier, Lily Long was the only one of my four I actually knew.

7) July 1955 - February 1956, 2740 N. Central Avenue, Apt. #5, Indianapolis, Indiana

HONEYMOON IN THE SOUTH

[22] Pat and John on their honeymoon

We honeymooned in Daytona Beach, Florida staying with Pat's grandparents, George and Laura Palmer. Pat and I drove our 1955 Ford convertible to get us there.

The car was white with a Tropical Rose interior trim. This one arrived at the dealer before the model I ordered, so I decided not to wait.

[22a] Honeymoon

[23] Pat Achor at St. Augustine, Florida

The following pictures were taken during our Florida honeymoon at Castillo de San Marcos, the oldest masonry fort in the continental United States located in St. Augustine, Florida.

Check the silver coated Ray Ban sunglasses.

[24] John Achor at St. Augustine, Florida

SETTLING INTO A FIRST HOME

Our first "home" as a married couple was a converted attic (third floor) of a house on Central Avenue in Indianapolis, across the street from the Zaring Theater. Pat's best friend, Dottie Hannah and her husband Bill lived on the second floor. It was one large room in the shape of a cross; high in the center with the ceiling sloping toward the outer walls.

We both continued working at American United Life Insurance Company. I interviewed for a number of jobs, but no one was interested in a guy with a military commitment hanging over his head. Guess they were right, six months later I was called to active duty in the U.S. Air Force to attend pilot training.

THE AIR FORCE – THE BEGINNING OF A TWENTY-YEAR ODYSSEY FOR PAT AND ME.

ORIENTATION TO MILITARY LIFE

Four of us from the Indianapolis area piled into one car and headed south. On my way to flying school, I spent about six weeks at Lackland AFB, near San Antonio Texas for orientation into the Air Force. We wore this Flight A, Tiger patch on ball caps. Pat joined me in Florida.

Tiger Flight Cap Emblem

A friend of mine was afraid he would be held over and miss his slot in flight school. While on troop-duty, a period assigned to a group of new enlistees, one of them knifed another trooper. My friend was worried he would have to stay and testify. The whole thing became moot when the knifer said it was all a big mistake saying, "I meant to cut another guy." With a confession in hand, my friend moved on with us—on time.

8) February 1956 - August 1956, 1901 Avenue "H" N.W., Winter Haven, Florida – Primary Flight Training

PRIMARY FLIGHT SCHOOL

[25] Piper J-3 Cub

The photo to the left shows the Piper J-3 Super Cub in its original paint colors.

This is a screenshot from X-Plane 11, a computer flight simulator. The only flying I do in retirement.

In Winter Haven, we lived in a "garage apartment" i.e., living over a garage. Much like the one my parents rented in Hattiesburg, Mississippi.

[25a] Pat at Winter Haven apartment

As I recall, we did not have use of the garage itself. In the following pictures of Pat and me, the stairs to our apartment are visible to the right.

[25b] John at Winter Haven apartment

My training class, 57-I, was divided between three different bases; Bartow AB (Air Base), Florida (at Winter Haven), Bainbridge AB, Georgia (Bainbridge) and Spence AB, Georgia (at Moultrie). My assignment was to Bartow, where we flew the PA-18 (a J-3 Piper Cub with a 90-horsepower engine, Wow!). and the North American T-6G (Texan). The transition to newer training aircraft at the other two bases was complete.

[25c] Author by the last PA-18 Cub (298) he flew at Bartow AB

The T-34 and T-28 were built with tricycle landing gear; third wheel under the nose. Our birds were tail-draggers; main landing gear and a tail wheel. Landings were tricky and when mastered, we claimed we were the last of the *real* Air Force pilots.

DOWN TO THE NITTY GRITTY

[25d] Bartow—Bulldog Flight patch

The flight training was operated by a civilian contract company. Along with the civilian flight instructors, a couple of Air Force pilots were assigned to the base to administer final check rides. I bet all the ground and air instructors were pilots going way back. Our squadron ops officer was a redhead named Manuchy (not sure the spelling is correct) with a booming voice. That voice earned him the nickname Moose. Sounding like the old sergeant on *Hill Street Blues,* he bellowed out, "For Gawd's sake, stay away from those props."

Robert C. Branson must have been the greatest instructor in the world; he got me through Primary Flight Training. Mr. Branson was a civilian instructor hired by the Air Force to herd us to graduation. Most of the instructors believed if they shouted enough, the student would eventually learn to fly. Mr. Branson was cool and calm, and one day after a major screw-up on my part, he said in an even tone, "Well, John, what do you suppose you did wrong?" When a student soloed, protocol called for giving your instructor a bottle of whiskey. Mr. Branson preferred cigars – rum-soaked Crooks as I recall. Mr. Branson was certainly a candidate the Bless Them All, hall of fame; thank you Sir.

The Cub was a simple plane to fly with little in the way of bells and whistles. Minimum instrumentation, two seats and a stick for each pilot. Our only checklist was memorized: GUMPS – Gas (Fullest tank), Undercarriage, Mixture (Full rich), Propeller (Full increase) and Switches/Seat belts. In the Cub, a couple were superfluous; since the gear was fixed (non-retractable) we didn't have to worry about that one and there were no switches to check and being unable to get out of the seat in the air, the seat belt was fastened from engine start—we basically did a GUMP. We didn't drop the "U" since future planes we would fly would have retractable landing gear.

We were allowed twelve hours in the Cub, also known as the Bamboo Bomber, to solo. Being slow on the draw, Mr. Branson turned me loose at a bit over ten hours; I completed my three takeoffs and full stop landings to qualify for a dunking in the large round watering trough.

One afternoon our civilian flight commander called me into his office; he asked me if I had buzzed the city this morning. Buzzing, flying very low was a serious offense; possibly being kicked out of training. We had two flying periods in the morning and two in the afternoon. He said a local resident took down the tail number of the plane. I had flown that plane during the second morning period and a classmate flew it in the first morning period.

Unfortuantely, the person reporting the incident didn't get the exact time. The reported time range spanned a bit of each of the morning periods. The other student was a 1st Lieutenant named ▉▉▉▉▉▉▉▉▉ (redacted). For the most part, the students were Brown Bars, 2nd Lieutenants, but there were three Silver Bars rated as navigators who were accepted into pilot training.

I don't know what ▉▉ told them, but I vehemantly denied buzzing anything, anywhere, anyhow. As far as I know, the incident was not pursued further. It would be more than fourteen years and half way round the world before our paths crossed again.

MOVIN' ON UP

[26] Bartow AB T-6G

Here is the original paint job on the T-6G. This is a bird stationed at Bartow AB and I may have flown this (035) tail number.

Below 2,000 feet, note the open front cockpit canopy. Back seat empty.

Now I was ready for what seemed to me to be the largest airplane in the world. During World War II, the T-6 served as an advanced trainer. It boasted flying capabilities and power near that of the Japanese Zero. My transition to the Texan was smoother than learning in the Cub. Keeping the front canopy open was standard for takeoffs and landings. All we needed was a white scarf trailing in the slip stream. Once airborne, we closed the canopy and it remained that way until entering the traffic pattern for landing.

[26a] Flight line, T-6Gs at Bartow AB, Florida – 1st in line (488) is last T-6G I flew

Afternoons in Florida provided sporty flying climbing and descending past the cumulus clouds building in the afternoon heat. Bartow AB consisted of three intersecting runways and an active runway was assigned based on the prevailing winds. That gave six possible landing approaches.

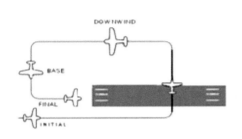
[27] T-6G Landing Pattern

New classes began to arrive and along with them new airplanes. We continued training in the T-6, but the next class would be flying T-34s and T-28s like the other Primary Flight Training bases. We now had three planes flying from Bartow each with a different landing pattern. T-6s approached 45 degrees to the runway heading, turn to align with the runway, initial, until midfield, turned 90 degrees and dropped about 300 feet to downwind, base leg and turned final around 600 feet above the ground.

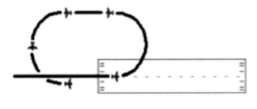
[28] T-28 Landing Pattern

T-28s followed the T-6 pattern until they reached the end of the runway; when the runway disappeared under the nose they executed a tight 180 degree turn to downwind followed by another 180-degree turn to final approach.

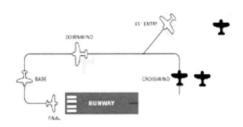
[29] T-34 Landing Pattern

The T-34 followed the more orthodox 45-degree entry to downwind.

With three different approaches to the same runway, the commanders were terrified of a mid-air accident in the traffic pattern. There were a number of solutions that could have solved the problem, e.g., everybody use the same pattern. Must have been something against that because their solution was to assign certain times for each pattern to be used. Each hour was divided into three segments of twenty-minutes each. I don't recall which segment was assigned to T-6s, but if you missed your landing period, there would be a forty-minute wait until you could come back in for a landing.

Missing your landing slot was frowned upon—by all, an woe to the student who screwed up the schedule for the rest of the day.

I didn't want to be that particular student, so I drew a diagram that would cover all contingencies. I created a cheat-sheet and put that drawing in the T-6G checklist we all carried. I've included the front and back covers of that checklist below.

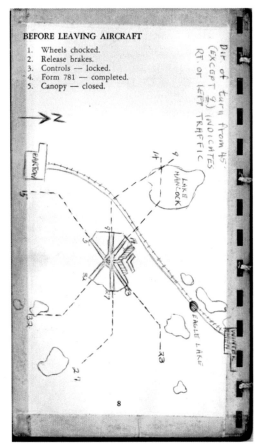

[29b] Inside back cover – 6 pattern entry diagrams

[29a] T-6G checklist cover

The checklist was orginally bound with flat plastic spiral comb. It didn't take long for the binder to break and disintgrate. I removed the broken plastic and replaced it with loops of cord passed through four spots on the left edge.

The brown border on the page edges is Scotch tape of the day. Especially noticeable on Figure 29b shows how it has discolored over time.

Runway headings are magnetic headings (001 through 360) shown as two digits. The three at Bartow were, 09/27, 14/32 and 05/23 — 09/27 was 090 (land east) and 270 (land west) which is an east/west runway.

A note on my traffic pattern diagrams reads: Dir[ection] of turn from 45° (except 9) indicates RT or LEFT traffic. Three runways support six traffic patterns where the prevailing wind allowed for little cross wind. Good thing, tail-draggers don't do well landing in crosswinds.

RISKING LIFE AND LIMB

After our 30-50 flight check, like a midterm exam, we were cleared for solo spins. Up to this point, we all accomplished dual spins, i.e., with an instructor in the back seat. To spin the T-6, we did a pair of clearing turns to be sure the sky near us was open; no other trainers nearby. We pulled the nose up 50-60 degrees and let the airspeed bleed off until we felt the bird stall; jammed full rudder and let the nose fall. The T-6 was a clean spinning airplane – meaning it put its nose down and rotated at a fairly stable rate. We picked out a spot on the ground and watched for the same spot to appear the first and second time completing a two-turn spin. Next, to recover we jammed in full opposite rudder (opposite the original turn in the spin) and popped the stick full forward. When the rotation stopped, we neutralized the rudders, added some power, and waited for the airspeed to increase indicating we were flying out of the spin. That was an important step, because hauling back on the stick could cause a stall and the bird would drop off into a secondary spin – not good.

Since we didn't carry oxygen, we were limited to a maximum of 10,000 feet. Above that altitude breathing became difficult and continued lack of oxygen can cause pilots to pass out. Well, I was cleared for solo spins and one day I convinced myself it was time to try one. I churned around the area for a few minutes screwing up courage. At last, I added climb power and headed for altitude so I could accomplish my first solo spin. I didn't want to exceed our altitude limit, but at the same time I wanted some extra altitude – just in case. I chose 9,800 feet as a good starting point. spin and successful recovery, took a deep breath and went for it. I stalled her, began the spin, counted two turns and initiated the recovery. I came out the bottom in fine shape thinking: I did it, I did it, Yee-haw. That was my one and only solo spin and I was embarrassed stopping at one. Later, in discussions with other classmates, I learned that many of them left Primary Flight Training with Zero spins to their credit. Since the Air Force didn't track those statistics, no one would ever know. But I knew and felt proud of myself.

Up until now we concentrated on the basics of flying the airplane. Now it was time to give cross country fights a test. After a dual day flight, we were turned loose on our own with the warning to stay below cloud decks—keep the ground in sight because we were flying dead reckoning.

I mentally reviewed all the steps for a (DR) and pilotage legs. DR required us to fly a prescribed heading for a calculated time at a specific airspeed and pilotage meant matching ground objects with our map. With a map and navigation log in hand, I launched my body into the wild blue.

Afternoon cumulus clouds were getting thick and closer to the ground. I looked ahead and decided I could get under them by flying at 1,000 feet. Since Florida is flat and low, MSL (mean sea level) and AGL (above ground level) were nearly the same. I neared my first turn point at Vero Beach airport. I reported to the instructors in their own planes circling the turn point; they didn't want anyone to get lost. Instructor said: I don't see you. We couldn't go farther unless the instructor had a visual on you and cleared you on the next leg. I repeated my position. Instructor: How can you be sure? I said: The name VERO BEACH is painted on the runway. When asked what altitude I was flying, I told them 1,000 feet. Apparently, they were above a near solid cloud deck. 'What are you doing down there?" the instructor said. "It's VFR (visual flight rules) down here." They sent me on my way and the remainder of the flight was a smooth one.

FACING THE TERRIFYING PART OF THE DAY—HOURS OF DARKNESS

The next adventure in flying was doing it at night. If daytime could be terrifying, what would it be at night? It's daaaaark, outside the cockpit. Interior lights for night flying wore red covers to keep from ruining night vision. Each instrument had one or two very small lights to allow the pilot to determine the instrument reading. We were required to have one solo night landing and a dual cross-country to graduate. For the night landing, staff assigned each student a number. We started engines and lined up on the taxiway. At "official" sunset—which is still pretty much daylight, the first plane in the stream took off.

I remember the launch interval as being three to four minutes with maybe ten or twelve in the stream. First to launch made the landing in day light—still officially a night landing. Last to land accomplished the feat of derring-do in twilight. I was somewhere in the middle and got the beast back onto terre firma without incident.

By now, a new instructor took over Mr. Branson's students. I don't recall his name, but he was affable and not a yeller. I still recall his philosophy for flying: the objective is to get from Point A to Point B without loss of life or government property. His briefing for our night cross country was: Achor, you f****r, don't get me lost, I'll be asleep.

GROUND SCHOOL

Our ground school instructors prepared us for most of the challenges we would face. All were flyers of one sort or another. One, with a arm permanently bent at a 90-degree angle told us the story behind the arm. During WW II he flew as a crop duster to boost the war effort. Being short of pilots to do the job he worked far too many hours. One day as he flew a few feet above the crop, he fell asleep. When he awoke, he was aimed at the boundary of the field, a line of trees. He yanked back on the stick but it was too late. He crashed into the trees and the bent arm was the result.

We were required to read and write Morse code; the dit's and dah's spelling out letters and numbers. On the first day of class, the instructor told us what we would be doing in the class all the time tapping the pencil in his hand on the desktop. A few minutes later, a couple of the students rose and left the class room. He told us the pencil tapping spelled out "if you can read this, you are excused from the class." By the end of the course we could read and write seven words per minute of aural sounds. We could also read four words per minute from a blinking light. There was no guessing is we missed a character; there were not any words; just random characters in groups of five.

Our navigation instructor gave tests nearly every day. My table partner and I raced each other to finish the daily project. We were pretty close to each other on being right or wrong; but the competition forced us to work hard and fast. The results became obvious when we both scored 100% on the final exam.

A physiology instructor armed us to face the scary parts of night flying. In a small room, he turned off the lights and the room was light-tight. It was pitch dark when he turned on a small, dim bulb mounted in the corner of the ceiling behind us. He asked us to watch it and let him know if it moved and if so, how far. Answers ranged from none to several inches. When he turned the room light back on we could see the light was a permanent fixture. We learned that staring at a stationary light in the dark will cause it to move. A perfect way to get vertigo at night.

We learned that all the parts in the eye that form vision find their way to the back of the eyeball and exit in one spot to the optic nerve—forming a blind spot. Daytime, not too bad, but at night can cause problems. Staring at a fifty-five drum at night at a distance of 50-60 feet (best as I can remember), the barrel will "disappear." Not good for avoiding taxi accidents at night. This was our first introduction to vertigo, which can mess up your mind. The ear contains three semi-circular canals which have tee-ninecy hairs growing inside. The canals are in three different planes and give us a sense of up and down and turning. If the fluid in one canal moves, no problem; fluid in two canals moving, you can still "feel" what's going on; but if all three canals are involved there is no way of telling which way is up. Not a desirable condition to face while flying a plane.

For more about this phase of my life, refer to the short story—*I Learned About Flying From That–The Piper Cub—PA-18* in the Appendix.

After graduating from Primary Pilot Training, we were our way to Texas, and we stopped in Gulf Shores, Alabama. It's on the east side of Mobile Bay. Pat's dad put us in touch with his brother, Dave Hendry and his wife Sug. The had a home in Gulf Shores and invited us to stay with them. They took us out to dinner and introduced a friend they knew. When the fellow left the table, Dave told us he was a millionaire and asked us to guess what he did for a living. We had no idea; Dave said he was in the junk business. There's money to be made everywhere. We would stay overnight with Dave and Sug a few years later on the other side of the U.S.

9) September 1956 - February 1957, 304 Johnson, Apt #9, Big Spring, Texas

MOVING FROM PROPS TO JETS

From Primary Flight Training, in Class 57-I, I moved on to Basic Flight Training. We split with part of the class going to multi-engine training in the B-25 Mitchell built by North American as a medium bomber in WW II. The rest of us headed for bases where they were flying single engine jet trainers.

[30] 2-ship formation - T-33s

Time for another move; this time to Big Spring, Texas – a town where the spring dried up decades before. Here's where bad weather began to follow us—this area at the bottom of the panhandle suffered from a drought for some years and the year we arrived they needed flood relief. We were about 30 miles east of Midland which became more famous as an oil center.

[31] Author in T-33 cockpit

Webb Air Force Base was on the west side of town and consisted of a number of military buildings with the runways on the west side. It had a pair of north-south runways each around 10,000 feet in length. Here we were introduced to the Lockheed, T-33A, Shooting Star, a single engine, two-seat trainer. The seats were tandem with the instructor occupying the rear seat for flights, except for instrument training. On those missions, students sat in the back and pulled a canvass "hood" over our heads so we couldn't see outside the plane forcing us to use instruments only to stay the course.

The plane was powered by an Allison J33-A-35 centrifugal compressor turbojet. Original prototypes, the XP-80 single seat jet fighter came from the Lockheed Skunk Works by Kelly Johnson. Johnson also produced the U-2 high altitude spy plane and the Blackbird, the Mach-three SR-71. Remember, this is the end of 1956 and the SR-71 was still a dream. However, on a cross country flight one turning point was over Laredo Air Force Base, Texas—on the U.S.-Mexico border. From the front seat, I looked down at the flight line and saw several peculiar looking airplanes: long slender fuselage and even longer wings that resembled a glider. I asked my instructor what they were. After a pause, he said, "We don't talk about them, understood?" I got the point, kept my mouth shut and later learned Laredo was a training base for U-2 pilots.

YOU NEVER KNOW

A note about two classmates I knew. One flew with a cold because he needed a small amount of flying time to complete all the checkmarks for graduation. His clogged sinuses blew out his ear drums and he was physically unqualified to fly. The powers that be decreed he would graduate with our class, receive his wings, and be permanently grounded at the same time.

The second flyer, graduated on time and as far as we knew he was doing fine. Then the word made the rounds; he took his own life.

Seems like the first person was more likely to have problems, but the reverse was true. You never know…

The centrifugal flow engine provided enough power, however when power was needed in a hurry, it came slow. We were admonished to never try to throttle-burst the engine, i.e., ram it to full power in a quick burst. Not only would full power not result, she would likely go into compressor stalls—the air passing the internal turbine blades would stall. We became adept and coming up to full power in slow smooth movements. On one of our first trips to the squadron building, a photographer snapped pictures of each of us clambering into the T-33. The ladder was wide enough to let both pilots get into the cockpits without moving the ladder fore or aft. Here I am wearing a flight suit and flying boots with the parachute hanging behind me and holding the helmet.

[32] John boarding a Shooting Star

[32a] Our WW II type pinup, by classmate Pete Winter (image was included in our class yearbook – The Bird and I ...)

The T-bird was easy going plane to fly. It was supposed to have smooth spin characteristics, but we never spun the bird; not even instructors. Instead, we learned to recognize situations that might cause a spin and avoid them. She was a straight wing plane which meant she was limited to sub-sonic flight. I was lucky again. The instructor I drew was William K. (Bill) Ayres; a man in the same mold as Mr. Branson. I'll take a moment here to thank him in advance—thank you Bill. He was a fighter pilot at heart and a short time after we graduated from Basic Flight Training, he was posted to Korea flying F-86s. Later he earned a Master's degree and taught English at the Air Force Academy.

We went through the drill of learning the plane in ground school and in the air and went on to solo. On solo flights we paid particular attention to the stick in the rear cockpit; didn't want anything to get wrapped around it and limit the front seat pilot's control of the plane. Weird things can happen and better to avoid them. Speaking of sticks in the back seat, the T-6's second control stick could be removed from it center position and stored on the left side of the back seat. This caused a bit of trouble for a student one day. He was in the back seat and the instructor told him to take over the airplane. The instructor turned loose of the stick and the bird fell off on one wing. Instructor yells at student; students says he's pulling as hard as he can but stick won't budge; instructor realizes the rear stick is stowed. Yep, weird things can happen.

FROM FUN (CONVERTIBLE) TO CONSERVATIVE (SEDAN)

[33] Our 1956 Chevrolet

Our home was on the second floor of an "apartment" building; don't think it started out as such. A closed in section of an outdoor porch became a kitchen. The wind blew in west Texas; didn't need to open a window to let the dust in. In anticipation of an enlarging family, we bought a new car. The 1956 Chevrolet 4-door was white over silver over white. Note the classy white sidewall tires. The gas filler cap was disguised behind the left tail light assembly. The picture shows the new automobile on a hill overlooking the city of Big Spring. Sorry about the quality; over the course of sixty some years, the colors have faded.

The T-Bird did not possess a steerable nose wheel and with a 90-degree free caster—45-degrees left and right it could cause trouble. Those with little skill using the brakes to alter the path of the plane on the ground found themselves with a "cocked" nose wheel. Once the wheel reached a left or right limit, it was tough to un-cock the gear. When it was cocked the pilot could enjoy taxiing in a tight circle; not a fun feeling. I started engines and when I began to taxi, I told Bill Ayres, my instructor, the brakes were a bit spongy—not reacting well to my brake inputs. Guess what? I cocked the nose wheel.

Ayres, took control and ran up the power working to uncock the nose gear. He was successful and continued down the taxiway. At the first turn he made, the nose wheel cocked itself again. This time, my instructor couldn't get it straight. It took a crew chief to help bounce the nose up and down to get the wheel straight again. I smiled under my oxygen mask and heard a comment over the intercom; something like "no smirking."

NOT ALWAYS "ALL'S WELL"

The first time I taxied to the north end of the east runway, I noticed a large black spot on the overrun portion of the runway. Back in the day, building installed a net to grab the landing gear if we were unable to stop the plane on the runway. The net stood about three feet tall perpendicular to the runway. Each end of the net was attached to an anchor chain on both sides of the runway. The chain was laid out in the direction of landing allowing a plane in the net to pick up and drag one link at a time. Good thinking.

The bird that created the large black, burn, spot was landing to the south. The net was in its lowered position, the stanchions and the net lying flat. The student landed short, on the overrun, and somehow managed to snag the net with the gear. The system doesn't work in reverse. The pilot was trying to drag the entire chain in one jerk. The chain snapped off a main landing gear dropping the tailpipe to the concrete. Before the engine was shut down, the black stain was forever burned into the concrete as a reminder: *don't never land short!*

For the better part of seven months of training, the word was *stay away from other airplanes.* And now, it was time for formation flying. We took off as Number 2 in a two-ship formation. My instructor neared lead's plane, told me to stay there and gave me control. Well, damn near no control. I was all over the sky, in and out, 100% power to zero and speed brakes to keep from running into the lead aircraft. I never was very good at fighter type formation but improved enough to qualify to the next segment of training.

Lieutenant. Ayres taught me a number of lessons that stood me in good stead for all of my flying career. One day I was in the back seat, under the hood, he told me he had the bird and to flex my right hand to relax. I was tense on the stick flying instruments and Bill sensed it. He told me to take the stick below the normal grip and use only my thumb and forefinger. It's hard to over squeeze the stick with two fingers. From that day on I could relax my grip on a stick or control wheel and do a better job of flying the bird.

Today, fighters have a zero-zero ejection seat. The pilot can safely eject with zero airspeed and zero altitude with a single movement—and survive. Back in my day, my first jet was a manual-manual affair. Normal sequence was first pull a handle to eject the canopy, the large plexiglass item that covered two pilot seats. Next came two seat handles; one to lock our seat belt and shoulder harness, and the second would eject our seat from the plane. Once out, we needed to manually unlock our seat belt and shoulder harness and kick free of the seat. If all that went well, we could reach for the D-ring on our left chest and deploy the parachute.

All those steps became automatic for the zero-zero systems. My instructor briefed his four students the first day on the flight line. "I'm the instructor, by the regs, I have to let you eject first. If we need to go, I will call "Bailout, Bailout, Bailout."—if you say 'huh,' you will be talking to an empty seat." Harsh, yes; but necessary. An instructor and student had a mid-air collision in the traffic pattern, 1,200 feet above the ground.

The instructor directed his student to bailout; then he ejected through the canopy—not a recommended practice but possible. The instructor separated from the plane and his seat, pulled his rip-cord, the chute opened, did about a half swing, and he touched down on the ground. He was the only pilot of the four involved who survived. Another lesson; sometimes you don't have much time to react to survive.

LIVING WITH OLD TECHNOLOGY

The fuel system on the Shooting Star was from WW II days. Again, today most planes have a single-point refueling system, i.e., plug the refueling hose into the plane and fill all the tanks in the bird. Not so for us. The crew chiefs caught the brunt of topping off all the tanks, but we had to check every filler cap on the bird. As I recall, that was five tanks, two leading edge tanks, two wing-tip tanks and one fuselage tank – open the lock on the cap, open the cap, check the cap was tight and reverse the process.

Missing the check on most of the tanks wasn't life or death; simple a loss of fuel until that particular tank was empty. The Fuselage tank was a horse of a different color. At low airspeeds, the engine could not get enough airflow to operate properly. The answer was to build a plenum chamber (a place for air) around the engine and allow air to enter via a pair of spring loaded plenum doors. Low pressure in the plenum chamber sucked the doors open (inward) at low speeds and when the airspeed built to a decent level the doors would close. The plenum doors were just behind the canopy. The cap for the Fuselage tank was between the canopy and the plenum doors. The cap, if not secured properly, would allow fuel to siphon from the tank and enter the plenum chamber. The place where all the fuel and air get mixed for combustion—and provide an extra Boom which might take the whole plane with it.

EXTRA FUEL IS GREAT, UNTIL …

The tip tanks could create problems as well. They were added to the plane to provide more time in the air. I found one reference relating to their capacity, indicating 230 gallons which translates to, at 6.5 pounds per gallon, almost 1,500 pounds of fuel hanging on each wing tip. We had no control of them because when on, there designed to both feed the same rate at the same time until empty—now, what could go wrong with this situation? I remember a warning light in the cockpit indicating no fuel flow from the tip tanks indicating they were empty.

With fuel in the tip tanks, some aerial maneuvers were restricted. Landing with fuel in those tanks was a no-no. Don't recall if this accident involved a dual or solo flight. The wing tanks refused to feed and the plane was recalled. There was a panic button at the bottom of the instrument panel set into a deep cup so it would not be depressed by accident. That button controlled a system that would jettison both tanks at the same time. Protocol called for the bird to approach the runways from north to south, lined up with the area between the runways at 500 feet above ground level. At midfield the tanks would be jettisoned.

The pilot followed procedures. Nearing mid-field, he pressed the panic button—only one tank left the airplane. All the weight from the remaining tank pulled downward on the plane and sent it into a cartwheel. The plane crashed, killing the occupant(s). Procedures remained the same, except the altitude was raised several thousand feet; giving a pilot a fighting chance for a bailout.

AIN'T NOTHING WRONG WITH BEING LUCKY

Sometimes we learned by surviving bad situations. At other times, we learned from the mistakes of others who did not live to tell the tale. We strove to be good at flying and if we were lucky as well, so much the better.

I flew enough hours in the backseat learning instrument flying; ignoring the feeling in the seat of the pants and putting my faith in the gages on the instrument panel. On one flight, I suffered a severe case of vertigo. Under the canvass hood, every part of my body said we were in all states of flight except straight and level. I told my instructor, Bill Ayres, how I felt. He said, "Okay, just keep fighting it and concentrate on and believe the instruments." Some of the best advice I ever got. I learned of could fight the physical sensations and fly out of it on the gages. Held me in good stead one dark night over the Artic and many other times. More of that one later.

Our apartment was on the second floor of an old office building, or something. What had been an outside porch was closed in and formed our dining room and kitchen. When the wind blew, which was a near daily occurrence, the dust or snow—depending on the season—penetrated the loose-fitting windows.

Pat experienced severe morning sickness. She would put breakfast on the table and make a dash to our bathroom.

EVER HEAR OF TESTITUS?

Bill put me up for my instrument check ride just before Christmas 1956. Later he told me it was to get it out of the way so I could enjoy the holiday with family. He knew me well. Unfortunately for me, I drew the terror of the squadron staff. I'm sure the guy was great, but he carried a rough exterior and we feared him. I settled into the back seat of the T-33, ran my checklists which included setting the low-frequency radio; a radio used the flying radio ranges (from the 1930s), RDF (range direction finding) and ADF (automatic direction finding). These stations used frequencies from around 200 kilocycles to about 400 kc – just below the band normal AM radio uses.

I was as smug as I could muster at this point; then I couldn't identify the radio station. I was tuned to something 100 kc off the desired number. After another minor screw-up, the check pilot said, "How many hours of instruments have you got, Ace." The sarcasm wasn't lost on me and from there it went downhill. Needless to say, I went on holiday leave with an instrument re-check haunting me.

I don't know why I choked up on the check ride. I was a good instrument pilot. My instructor taught me how and even coached me through basic aerobatics while flying under the hood. All that prep and I clanked.

In January, I overcame my "testitis" and passed my instrument check and the final check ride. Bill Ayres pinned my wings on my uniform and we graduated as full-fledged pilots in the USAF.

FUDGING THE RULES

During training, a requirement for graduation was a night wing formation takeoff. We were scheduled as Number 4 in a flight of four; Lead would take off with Number 2 on his right wing giving that student his wing takeoff. As Number Four, I would be on Three's wing to complete the requirement. On the taxi to the runway, our element lead, Number 3 called an abort. Ayres, feeling the whole night falling apart as a waste of time, called the tower.

He asked that we be able to join Lead and Two as a three-ship night formation flight. Lead was waiting in the center of the runway with Two on his right wing. Without mentioning it to the tower, Ayres told me to line up on Lead's left wing. We ran up engines to max power and Lead called for us to release brakes. We were on our way, I could get my night wing takeoff, and the tower didn't realize what was going on until we were racing down the runway. I seriously doubt that three-ship takeoffs were authorized. That was the night l learned one of a couple of precepts I followed from then on: *If you can't afford a No answer, don't ask the question.*

AT TIMES, LUCK ISN'T ENOUGH

Sometime during our time in Big Spring, Pat and I went to their drive-in movie. Cosdon Oil Refinery, a sprawling complex was north of town and behind us as we watched the movie. For some reason, I looked out the rear window of our car and saw a giant flash near the refinery. A Webb crew-chief apparently decided that if these dumb brown-bars (2nd Lt) could fly the T-33 then he could too. Crew chiefs were qualified to start engines and taxi birds around the flight line; to maintenance hangers, etc. That night, the crew chief headed for the runway and launched into the night sky.

The crew chief used the radio to advise the tower of his situation. The tower personnel were doing their best to talk him down. One instructor told him to look down in the cockpit—for a gauge or switch—and he never looked up.

Unfortunately, he wasn't able to keep the bird airborne, but fortunately missed the main Cosdon complex, and he smashed into the ground just outside the perimeter fence and perished in the crash.

FIGHTER JOCKS

Near the end of training, several of us were on the flight line looking at an F-104; the first plane which had more thrust than weight. A young fighter jock arrived to preflight his bird and asked us if we wanted to see a vertical takeoff. Of course, we said yes. The jock told us to hang loose for ten minutes or so. In that time, he cranked the only engine, taxied to the runway for takeoff. As we watched, with full burner going, he retracted gear, flaps and raised the nose. When he was pointed straight up, he just kept on going—leaving our bunch of watchers, slack jawed.

10) February 1957 - July 1957, 8 E. "C" Avenue, Glendale, Arizona

GUNNERY SCHOOL

The life of vagabond Air Force families was put to the test again, so off to Advanced Flight Training in Arizona. Glendale and Luke Air Force Base were on the north west side of Phoenix.

[34] F-84F Thunderstreak

In those days, driving out of Phoenix, we could smell the onion fields. By now they are no doubt paved over. We rented a small apartment in Glendale, Arizona.

I was here to fly the Republic F-84F, Thunderstreak. She was a typical Republic bird; overweight and underpowered. She was the forerunner to Republic's F-105 of Vietnam era fame; whose nickname was Thunderchief; but was affectionately called the Thunder Thud.

Rumor said the F-84F was scheduled to have the Pratt & Whitney F-100 afterburner engine; but the airframe was ready before the engine and we wound up with a J-65—less power and no afterburner. The F-84s before this model were straight wing versions and therefore sub-sonic. The "F" model was a bent-wing bird, meaning the wings were swept back and was capable of supersonic flight … yeah, right! More later.

We began ground school, learning about this jet fighter which sported six .50 caliber machine guns fixed in the nose and firing forward. A .50 cal. round is a half-inch in diameter and leaves a nasty hole in things it hits. Since the F-84F only came in a single seat version, our poor instructors had to stand on the wing and lean into the cockpit during our taxi training sessions. Don't think we lost any instructors that way.

Our first actual flight put our instructor as Number 2, i.e., flying on the students wing so he could keep an eye on us. Got up, got it back down. Back in T-33s the ailerons were hydraulically boosted to give us control with high speed air over the wings. We learned to use a light touch and not over-control the ailerons. The ailerons in the F-84F were boosted as well as the elevators. Even with warnings from instructors, we tended to try to shake the tail off as we overcontrolled the fore / aft stick movement. Didn't take long to compensate and be smooth on the stick. With those boosted controls, she was a quick and responsive airplane.

MACH BUSTER—BARELY …

About our third ride, we got to go supersonic. Following my instructors briefing, I kept the Mach indicator as high as possible during our climb to 35,000 feet. Reaching altitude, I let the nose fall while maintaining climb power. When the nose fell through the horizon, I began shoving the stick forward until we were around 60 to 70 degrees nose down. I held that attitude as we screamed toward earth with one eye on the Mach meter. It crept up and up; reached 0.9 Mach, 0.95 Mach, and finally read: 1.01. A grunt over Mach 1 – the speed of sound.

At that point, recovery consisted of reducing power and bringing the nose up to bleed off airspeed and attain straight and level flight. Oh, well ... it was supersonic and we did it and could brag.

INDIVIDUALISM IS NOT ALLOWED

During the landing pattern, we gave a mandatory call on base leg to indicate that the gear was down and the brakes were checked. Tapping the brake pedals caused the hydraulic pressure gauge to fluctuate showing that the brakes should work after landing. The radio call was: [callsign] base, three gear and brakes checked. A classmate got bored with this standard call and one day turning base leg, he called: [callsign] three rollers and two binders. Word came down; there would be no further individual base calls.

On one solo flight I missed a step after retracting the flaps on takeoff. The flap switch was a three-position affair; Up, Off, Down. After I was airborne, I pushed the switch to the Up position and forgot the return it to the Off (center) position. Back in the landing pattern, I put the gear down and flicked the flap switch one position aft. I thought it was in the down position, but one click put it in the Off position. I started my turn to final and felt a light burble. I thank my dozens of stall recovery exercises going back as far as the T-6. Within a half-second, I rolled wings level, to increase lift and rammed the throttle to 100%. I recovered from the impending stall, but my actions required that I execute a go-around and enter the pattern for a second approach.

I explained to my instructor what happed and admitted to the screw up. I could have lied and said I just didn't like the look of the approach and went around Don't remember if that was my first pink-slip – a grade sheet indicating UNSAT, a flunked ride—or one in a line of them.

I do remember my instructor taking me outside for a serious *discussion*. I can remember the hot sun, squatting in the shade of the squadron building, smoking a cigarette. There was a serious tone in my instructor's voice as he spoke. I won't repeat the exact words because they cut deeply; the essence of his words was that I should quit the program on my own; that I couldn't cut it. Back in the day we called it an SIE – self initiated elimination. He shook my confidence and for a moment I thought it would be the easy way out.

That feeling lasted only a moment. I don't remember what I said to him, but my decision was to prove him wrong. I stayed in the program and fought to retain my wings.

Looking back, I think I know the reason for his request. I'm sure the squadron and instructors were evaluated on the number of students who graduated from the flight school. My best hunch is that tossing someone out of the program was a black mark on their report card. However, if they could convince a student to quit on their own, there would be no stain on their record.

THE UPDATE I MENTION IN THE FORWARD

My instructor had his say; I had my thoughts. In retrospect, I think that day his words steeled my resolve. I would suck it up and become the best pilot I could be. There were times I was a slow learner, but I did learn well storing away those nuggets of wisdom for future use.

Years after the Air Force, I was a real estate trainer and our national trainer used Practice, Drill and Rehearse as a mantra. I used the story of my accident in Rivet Ball as an example of Practice, Drill and Rehearse. Real estate might not be a literal life and death situation, but figuratively ignoring the mantra could easily be the financial death of a career.

Without being aware of the mantra, my approach was the same. I studied, worked to be the best pilot I could be and planned ahead to be ready for the unimagined. I approached my flying duties with an air of paranoia—who is out to kill me today and how will they try it? You can always hope for the best, but if you aren't prepared for the worse, you may not survive *that* day.

Lest you think I was dwelling on the negative, I feel it was just the opposite. I don't believe any of us who flew would have continued if we got up in the morning thinking today is my last one. With preparation, I could get up each day thinking it's just another day and I'm ready for it.

When *that* day happens, and there's a good chance it will, there are two types of approaches to the situation. Without practice, your actions can exacerbate the problem and make it worse or deadly. With Practice, Drill and Rehearse your actions are more likely to even out the problem with positive results.

WHY AREN'T THEY AROUND WHEN YOU WANT THEM?

One day, while running my pre-engine start checklist, I found the fuel supply valve would not open. If it ain't open, the engine won't start. We'd been warned of this in ground school; on shut-down, let the engine totally spool down before turning off that valve. Failing to wait would create a vacuum-lock in the fuel line and the "padlock" that fuel supply valve. There I sat in the cockpit, the victim of the student before me being too impatient to wait for the engine to stop rotating.

I could wimp out and tell my lead I couldn't get an engine start, or...I remembered a ground school idea that might work. We started the bird with an explosive charge that gave the engine a spin to assist in the start. We only had one shot. I don't remember all the steps, but I needed several near simultaneous actions to accomplish the start this way. I signaled the ground crew, fired the start cartridge, spun the engine, forced the fuel valve open and brought the throttle up to idle. Lucky as hell, she fired off and I could complete our mission. Where the hell are instructors when a bad situation goes well? I'm the only one who knows what happened. Reporting it, might have put the previous student in a bad light.

FORMATION WHAT?

Formation flying was still my weakest subject. Following a sloppy day flying on my instructor's wing, he gave me a pink slip—flunked the ride. He also put me up for a check ride with another instructor without giving a reason. After takeoff, I slid in close on lead's wing and vowed I would stick to him like glue. I was close enough to count the rivets in his aircraft's skin. I soaked my flight suit with sweat, but stayed with him for the entire ride.

This second instructor gave me a pink slip for flying too close. Arrrrgh! Two in a row was usually a death knell. My instructor went to him pleading my case – pink slip last time out was for sloppy formation – cut him some slack. He relented and revoked the pink slip. My career was still alive.

CATCH ME IF YOU CAN

Another lesson learned; blackouts. I saw from the flight schedule I would be going on a in-trail acrobatic mission. Four planes, one behind the other flying loops and rolls and anything else the leader could think of. The bird behind lead, Two could bobble a bit and then Three would wobble more and finally Four become the tip of the bull-whip. I was going to be Tail End Charlie, number Four.

We went through several maneuvers and I was holding my own but had allowed a gap to develop between my plane and the one in front of me. I was playing catch up. Lead announced a loop and began a dive to pick up speed. I pushed my throttle up in an effort to close the gap with Three. The instructor began his pullup earlier than I expected and I had to pull a lot of back stick to stay near. The G-forces were building. My G-suit, designed to inflate and stave off the effects of high G maneuvers, was working overtime. I was grunting and tensing stomach muscles to assist the G-suit.

I had to pull back harder to even come close and the harder I pulled the more G's I created. The effects of the G forces began to drain the blood from my head. My vision began to dim, then I lost peripheral vision and I was looking through a narrow tube – aptly named tunnel vision. I eased off on the power and continued to pull G's and the diameter of the tunnel narrowed—finally the tunnel closed and I was in a blackout. I was not unconscious, but I could no longer see.

At this point, my only option was to reduce power and ease the stick forward. As I did, the tunnel process reversed itself and I could see again. The other three planes in my formation were nowhere in sight. I was sent home alone, but I learned a valuable lesson that day.

ROOTIN' TOOTIN, SHOOT 'EM UP

Gunnery was our next phase, air-to-ground, air-to-air and bombing. We even practiced nuclear weapons delivery in the form of toss bombing and over-the-shoulder tactics. On that latter one, we drove into the target, pulled up and released our 100-pound practice bomb going straight up. After the release, we pulled over the top, headed for the ground at full power to get out of the blast area. Skip bombing was exhilarating, inbound to the target, fifty-feet off the deck around 250 knots kept my attention. With the proper gunsight picture, I toggled off the bomb; a good run was a one-skip on the ground and into the target. This was probably the best weapons delivery style for me.

The F-84F carried four .50 cal. machine guns—however, to conserve ammunition only two were loaded for students. Air-to-ground gunnery put four of us in a pattern taking turns at 20-foot square white canvass targets. The turn to final heading to the target required a radio call: right, white and 80. Right indicated the gun switches were set; white meant we had the target in sight and power was set to 80%. Coming down on the target at higher or lower power settings screwed up the ballistics for the run. I made the calls okay, but my targets seldom needed much repair meaning I didn't put many holes in them.

My prowess at air-to-air gunnery was a bit worse than with those ground targets. Two instructors flew a T-33 and towed a white sleeve or "banner" which was our target. We approached the target flying at 20,000 feet from its six-o'clock position, offset left, with the four-ship formation, in-trail and a few seconds between each bird. Abeam the target plane, execute a hard-right climbing break of 90-degrees. Half-way to 25,000 feet, we began a left 90-degree turn to a perch position. We now paralleled the target heading. From the perch, we rolled in, left turn and then right to set up the gunnery pass.

Another problem flying this air-to-air pattern was a lack of power in the F-84F. Climbing the five-thousand feet from abeam the target plane to the perch, was a struggle. Being used to that, it caused me a problem when I moved on to advanced gunnery at Nellis, AFB, Nevada.

The T-Bird pilots stretched their necks to be sure the students didn't make a "flat pass" which would mean we might be aiming at them and not the sleeve. My major problem for air-to-air was an inability to range the focus of my eyes outside the cockpit. I didn't "see" the target sleeve early enough to fly a good pattern. Not sure I ever hit that banner with a .50 cal. round.

THROWING BOMBS EVERY WHICH WAY

We learned nuclear weapons delivery as well. Two types of delivery included toss bombing and over-the-shoulder. To toss the weapon, we hit an IP (initial point), began a pull up and released the practice bomb—a 100-pound blue bomb, about three feet long—somewhere around a forty-five degree nose up attitude. From there, we executed a hard-breaking turn and hauled ass away from the target—which is what you need to do if you're tossing an actual nuke.

Over-the-shoulder bombing moved the IP to the target itself. We bored in and began a hard pull up directly over the target. When we reached a vertical attitude, we toggled the bomb off. After release, we pulled hard, over the top bringing the nose to a forty-five-degree dive and since we were then inverted, rolled upright. I think I came close to the target on some of these last two types of bombing.

The third bomb release we learned was the skip-bombing run. We bored in at the target in level flight about fifty feet off the ground at somewhere around 250 knots. When we had a proper sight picture, we again toggled the bomb off the wing and began a pull up. The weapon was supposed to do a one skip off the ground and strike the target. This was the one where I could hit the target on a consistent basis.

And so, struggle as I did, I graduated from gunnery school and moved on to my next assignment—flying the F-100A at Nellis.

11) June 1957 - August 1957, BOQ, Nellis Air Force Base, Nevada

ADVANCED GUNNERY SCHOOL

Pat was pregnant with our first child and we decided she would stay in Phoenix. We had the apartment and she liked the hospital facility and her doctor. The greatest saving grace—her mother came to stay with her in Phoenix.

[35] F-100A model

I was consigned to the luxurious accommodations in the BOQ (Bachelor Officers Quarter) on base at Nellis. We got an orientation ride in a T-33A with our instructor and moved on to the F-100A—a single seat fighter which was the first USAF fighter that could attain the speed of sound (Mach 1) in level flight with an assist of the afterburner of course. This plane carried more punch that my last one; it sported four 20mm cannons.

One of our early flights included a brief time above Mach 1, super-sonic. North American Aviation sent us a mach busters pin and certificate. Figure 36a shows the pin on a cap I received when I was included in a 2017 Honor Flight to Washington D.C. while in Nebraska.

[36a] Honor Flight Cap with F-100 lapel pin

[36b] Mach Buster Certificate

The certificate from the plane builder in shown in Figure 36b.

The usual ground school courses ensued teaching us about the plane and we were expected, like the single-seat F-84F, we lugged this after-burning beast into the air all by our lonesome. This plane again had boosted ailerons, rudder and horizontal stabilator. The one thing it did lack was flaps, which give better lift at slow speeds and allow lower landing speeds. Our final approach was around 180 knots with a touch down speed of 160 knots (about 185 mph). That's moving right along.

It also sported a speed brake used to kill speed in the air and on landing. It was a huge slab which extended from the belly and swept forward around seventy degrees. Like the T-33A, a good amount of pitch up was caused by extending the speed brake—forward stick and pitch trim was needed to maintain level flight.

Another extendable we had was a tail skid on the aft end of the fuselage. It was intended to protect the after-burner area in the birds' tail cone. We didn't have to remember it, extension and retraction was automatic, tied to the gear position. This was also the first Air Force plane to incorporate titanium. The afterburner area was protected with this metal since it could withstand the excessive heat created when the after burner was used.

FAMILY MATTERS

Pat stayed in our apartment in Phoenix during this stint. Her mother came down from Indianapolis to lend a hand. I drove the round-trip to Phoenix every weekend while I was at Nellis. One weekend in June, I was getting ready for the drive back to Las Vegas on Sunday. Pat said she felt "funny" but dismissed it. Back at Nellis AFB, my mother-in-law called to tell me I had a brand-new daughter, Kathleen. I reported to my squadron the next morning and told them about events.

I was on the flying schedule that morning; response: fly the mission, then take a day off to see my wife. So many squares to check and limited time to do it. Better than nothing.

STAYING COOL

The desert in summer was brutal and our '56 Chevy didn't have air-conditioning. With very low humidity, swamp coolers (evaporative coolers) work well. In fact, our apartment in Phoenix used this method of cooling. The unit is lined with material that conducted water well; water ran down the material and a fan forced cool air out of the unit. Even at 80+ degree temps, a swamp cooler could drop the house temperature 15 to 20 degrees.

I shopped around and located a small unit that would sit in the middle of the front floor of the car. It was powered through the cigarette lighter. The cooler impressed more than one person who hooked a ride with me to Phoenix and back.

PANIC ON THE ROAD

The highway to Phoenix was all desert. Just south of Las Vegas was Kingman, Arizona; just north of Phoenix was Wickenburg—in between those towns was more than a hundred miles of mostly nothing. I liked to gas up in Kingman southbound and in Wickenburg northbound—so I could run that road between the towns on a full tank of gas.

One night heading southbound, I forgot the fill the tank in Kingman. Well, no sweat there was an alternative. The town of Wikieup was about the mid-point on that hundred-mile dash and I remembered it had a small service station and maybe three or four other buildings. It was around ten p.m. as I roared southbound. Next thing I knew, I saw the outskirts of Wickenburg. What happened to Wikieup? They roll up the sidewalks early and I went through the town and never saw a light. Damn, glad I had fuel enough to make it.

This one happened in daylight on that same stretch of road. I was cruising northbound heading for Las Vegas somewhat north of the speed limit. In this stretch of desert, there were few obstructions on the road ahead, but there were a few. I topped a slight rise and what to my wandering eyes should appear; a herd of cattle covering the ground on both sides of the road and both lanes of the highway. Did I mention this was free range territory and there were no fences as far as the eye could see.

I was too close to even think of hitting the brakes; I'd never stop before hitting the first cow. At the last minute, I spied an opening wide enough to squeeze through, and the pedal went to the metal. I figured my best bet was to get through those bovines before they knew I was there. Sheer luck guided my hand that day; the car and I survived without touching anything but asphalt. I also learned what the letdown after an adrenaline rush feels like.

OTHER ODD MOMENTS FLYING THE F-100

I described the air-to-air gunnery pattern when I was flying the F-84F. getting to the perch, 5,000 feet above the target altitude (20,000 feet) was a struggle. I forgot I was flying an F-100 as I broke right over the target bird, turned ninety-degrees; rolled out and looked over my shoulder the locate my target and turn left to parallel the targets direction. The scan took too long. By the time I glanced back inside the cockpit the altimeter read: 30,000. I lost sight of the entire flight and got to go home all by my lonesome.

The next incident happened on another air-to-air mission, which meant each plane was loaded with 20 mm canon rounds. I was number Four in a four-ship formation. Lead, number one, was the instructor with a student as number Two on his wing. Element lead, number Three was another student and I was on his right wing. We launched in pairs and the two the elements joined in the air after takeoff.

Lead and number Two, who was on Lead's left wing, released brakes, lit their burners, and accelerated down the runway. My element lead, number Three, signaled to me and we revved our engines to full power and waited for the proper takeoff interval. Time's up; number Three signaled me to release brakes and then go to afterburner. I watched number Three's nose wheel and as it came off the ground, I began my rotate for takeoff. 'Bout that time the fit hit the shan.

Sometime before this, the lead instructor signaled an abort. Don't know what the problem was, but he and number Two chopped power and began braking. Guess he forgot he had two more birds close behind him and he didn't call the abort over the radio. By the time he and the tower realized the situation and began screaming "Abort," it was too late.

Number Three and I were climbing up his ass and were already rotating for takeoff. We came off the runway and probably scorched the lead birds with our afterburners as we screamed over them and into the wild blue. That's when the tower came unglued, because they had a two-ship of students in the air loaded for bear with air-to-air munitions—Hot guns, Oh, my.

The tower told us, me and my lead, to make sure our weapons were set on Safe; that we were to break up our element, burn off fuel and return to base as individual planes. I spent a half hour or so in the acrobatic area putting the bird through its paces and then headed home.

That day I learned that even instructors and the best pilots can screw up and put others in danger. It may have been that day I became a paranoid—plan for the worst; hope for the best; and be ready to handle whatever comes my way. I think that decision helped me stay alive and be ready for those occasions when the whole day goes to hell.

I graduated from gunnery school, and we headed back to Indianapolis for a short leave. We would spend time with our parents and show off their new granddaughter. August is not the time of year to travel with only a swamp cooler in the car.

Our daughter was about six weeks old when we began this odyssey. To avoid as much heat as possible, we decided to drive at night and stay over in an air-conditioned motel during the day. For me, it was a decent schedule, but it wasn't that easy on Pat. While I slept, she cared for our daughter and prepared formula for the next day using a small unit plugged into a wall socket. We'd pack up the car at sundown. Our daughter usually slept well in the car, but Pat did her best to stay awake to keep me company. She was tougher than I.

A couple of weeks later, the drive south to Texas was smoother. I think we reverted back to daytime driving.

12) September 1957 - January 1959, 3208A Oaklawn, Victoria, Texas

[35a] 452nd FDS patch

The day we arrived in Victoria, the local newspaper carried a front-page ad stating that Foster, AFB would be closing shortly. As it turned out, the base remained open for nearly a year and a half.

I was assigned to the 452nd Fighter Day Squadron whose hierarchy read: 322nd Fighter Day Group, 450th Fighter Day Wing, (under Twelfth Air Force at James Connelly AFB, Waco, Texas. The 452nd was the great grandchild of the 452nd Bomb Group, a WW II B-17 unit.

The squadron was assigned F-100C models. A bit of an improvement over the "A" model I flew at Nellis AFB, but not much—the bird still did not have flaps. One of my first tasks was to observe a half-dozen landings from mobile control.

Mobile control was a mini-control tower on wheels placed and the landing end of the runway. A qualified F-100C pilot was stationed in the mobile unit to grade landings. On my first visit I watched a couple of planes make good landings and then—the third one made a three-point landing; two main wheels and the tail skid. Earlier I described that skid as an addition to protect against dragging the expensive tail area and the afterburner.

In this case, the skid didn't protect anything because it was driven up into the tail of the bird. A rather expensive repair job. Next on my task list was a chase ride. I would act as the lead aircraft with an instructor pilot flying my wing. Everything went well up to brake release.

[36] F-100C in 452nd FDS green

I nodded my head, released the brakes and shoved the throttle outboard to kick in the afterburner. I felt the burner light and figured all was well, until the instructor yelled something over the radio; something about the afterburner. Being well past an abort point, I continued the takeoff and climbed as fast as possible to 1,000 feet AGL, came out of afterburner and looked around for the problem. The F-100C didn't have the most sophisticated ejection system—the thousand feet would allow me a safe ejection.

My chase pilot was up on my wing by this time and he told me about the problem on takeoff. Normal afterburner operation put out so much power and exhaust, the eyelids opened to increase the diameter of the tailpipe. The instructor told me my eyelids did not open and that generated excess thrust—he couldn't keep up with me until I leveled off and came out of afterburner.

We flew around for a few minutes burning off fuel then entered the traffic pattern and landed without further problems. Don't know what the repair bill was for that bird.

FROM FLYING JETS TO FLYING A DESK

Higher headquarters began slotting pilots for new assignments; using FIFO—first in, first out—the old-time pilots got first choice. Being tail end Charlie, it was obvious I would not get a flying position. There was an open billet in the Base Procurement Office, and I *scored* the job as Contracting Officer.

Base level Contracting Officers oversaw all purchases at the base while remaining within a $200,000 limit per transaction—that is the equivalent of more than $1.5 million in today's dollars. The only stumbling block to being assigned to the position was a requirement that any Contracting Officer must first complete the appropriate school before taking charge. Again, the powers, in their infinite wisdom, solved the problem—waive the requirement.

To the best of my knowledge, I was the only officer appointed a Contracting Officer without benefit of the required training. And so, I took over an office consisting of a civilian deputy, two male NCO buyers, two female civilian buyers, and a female receptionist.

In nothing flat, the deputy was transferred leaving me to OJT (on-the-job-training) mostly on my own. The scariest part was dealing with allocated funds which was a touchy subject. For the most part, you didn't get a second bite of the apple—get it right or go home.

During this phase of my tour, the base retained a couple of T-33s, so desk pilots could log enough flying time to qualify monthly for flight pay. Base Operations established the flying schedule, two pilots at a time for each bird. The first person listed for a plane took the front seat while the other pilot ended up in the back seat. I'd been second on the list for several flights in a row, so the day I was first on the list I climbed into the front seat. The major—chief of the finance office—seemed to grumble as he settled into the back seat.

After a couple of hours in the air, we entered the traffic pattern for landing. The major said he would show me how to do it—at the pitch out point he racked it into a tight 180-degree turn and yanked the throttle to idle. The challenge was to gage the pattern so he could make the second 180-degree turn, drop gear and flaps, and touchdown without taking the throttle out of idle.

His judgement was way off. Long before the end of the runway, he had to add a batch of power to keep from landing in the dirt. He was pissed I took the front seat without asking and embarrassed himself on the landing. We didn't exchange a single word for the rest of the period. That set the stage; more about the major later.

We scrounged flying time in anything that was available. Logged 30 minutes in this helicopter and a couple of hours in the amphibian.

Above, on the left is the UH-1 Chickasaw [37] and the SA-16 Albatross [38] — designated HU-16B in 1962—is on the right.

LIGHTWEIGHT AND DANGEROUS

The base had an inventory of lightweight .38 cal. snub-nose revolvers. Most likely built by Smith & Wesson, this gun was designed to be carried by aircrew members and aluminum was used for the frame and cylinder to reduce the weight. It was also used for annual pistol qualification.

At some point, it came to our attention that this weapon had a specific life span based on the number of times it was fired. That fact was not passed to the folks who stored and cared for the guns. No one had any idea how many times a given gun had rounds pumped through it. A failure of the cylinder could cause catastrophic injuries and the USAF decided to scrap the entire inventory of these guns.

ANNUAL CHECKRIDE

Foster AFB wasn't designed as an all-weather base. After all we were a fighter DAY squadron, apparently indicating we didn't like to fly at night or in bad weather. For instrument approaches, we had a single aid; an old ADF (automatic direction finding) beacon. In the plane we could tune to this station and a needle in the cockpit would point to it. Rudimentary and basic it could get you down below the clouds if they weren't too low.

The takeoff and climb were no problem and the clouds were only a few thousand feet thick. I was in the back seat, under a hood simulating total lack of visibility going through my paces to complete this instrument flying check. In clear skies above the clouds, I started through a series of maneuvers as the check pilot called for them. Then it got dicey.

Fairchild tower called and told us the ADF station had just gone off the air. There was no way we could return to our home base. We had enough fuel to go elsewhere, but that would dump more junk in the water. The tower called back and said he could lead us through a DF penetration. I referred to an ADF approach as rudimentary; DF was more like stone age navigation.

To locate us, the tower would ask for a short count—we would hold the mike button down and count from one to five and then from five back down to one. During that transmission, the tower would rotate an antenna to point at us and then give us a heading to fly. To stay current, the tower was asking for the count to be repeated often. He guided us back overhead at 20,000 feet, and from there we executed a jet penetration mimicking the existing approach—descending, turning back toward the runway and ending below the clouds. Fortunately, the cloud base was around 1,000 feet and we broke out of the clouds with the runway in sight.

On the ground, I said I knew I hadn't completed all the scheduled items for the check and asked when we could finish the ride. His answer was something like this, "Anyone who can fly a DF penetration in real weather has just passed his check ride as far as I'm concerned." Damn, I'm good.

T-BIRDS ON LOAN

Out of the goodness of their heart, our next higher headquarters, Twelfth Air Force Headquarters, located at Connelly AFB near Waco Texas, loaned us airplanes so we could get enough flying time to qualify for flight pay. They brought four T-33s down to Foster AFB on a Friday afternoon. The plan was for eight of us to take them on cross country flights and log at least two hours on Saturday and two hours on Sunday. The birds were scheduled to be ferried back to Waco on Monday.

The two pilots in one of the T-33s bailed out over the Louisana swamps and the plane was lost. Both pilots were okay. Another plane had problems of some sort and didn't make it back to Foster for weeks. One (out of the four) did make it home on time; and then there was the one I was flying.

The other pilot with me and I agreed to make Kirtland AFB at Albuquerque, New Mexico our destination for an RON. After a "remain overnight," we would fly back to Texas. Ooops! On our let down to Kirtland, one of the main landing gear dropped out of the wheel well. We notified the tower, lowered the landing gear and executed a fly-by so the tower could check our gear. We were getting an "down and locked" cockpit indication and the tower said all three gear appeared to be down.

We landed with no problem, and a maintenance team inserted landing gear safety pins which should prevent the struts from collapsing. Taxi in and parking went fine. To clear the write-up the plane would have to undergo a retraction test. Simple; put the plane up on jacks, raise and lower the gear to see if it operates the way it should. Now for the bad news: the equipment and personnel required to perform the test were not available on weekends.

We made it to the ramp early Monday morning. The gear test was complete and all was okay. The other pilot and I headed into Base Ops for a weather briefing. We wanted to file an IFR (instrument flight rules) — "1,000 feet on top." That would let us fly instruments until we were 1,000 feet above any clouds and maintain VFR (visual flight rules) for the leg home. The advantage here was flying a plane without an autopilot was easier when VFR. IFR requires more precise control and more sweat.

In answer to our request, the weather briefer said; "We got thunderstorms in the area that top out above 63,000 feet. Can you get that high?" He knew as well as we did, the T-33, on its best day, could not get anywhere near that altitude. He gave us an idea of where the storms were the lightest, and we re-filed our flight plan for a hard altitude we could attain. We finally made it home late on Monday and Twelfth Air Force never loaned us airplanes again.

This was also a period when I introduced myself to two-wheel motorized transportation. Since we only had one car, the '56 Chevy sedan, Pat was left without wheels when I went to work. Somewhere—don't remember where—I came upon an Allstate scooter—one seat, floorboard area for feet and a bicycle type steering handle and throttle. The back, under the seat and around the engine was sheet metal. I fashioned a couple of "saddlebags" from aluminum and riveted them to each side. The old engine probably weighed in around 10-horsepower, so she was limited to thirty miles per hour or so. By riding the paved shoulder, I managed the highway from Victoria to the base on sunny days. Now, Pat could go out with our daughter, Kathy, driving a more conventional mode of transportation.

ADDITIONAL DUTIES

As the chief of the Procurement Office, I also had an additional duty—Mortuary Officer. My duties involved coordinating with a local funeral home, viewing the body and helping to resolve any problems that might arise.

Fortunately, during my tenure, there weren't many deaths related to the base. There was one which was notable; a young airman died in an automobile accident. He was Greek, young, blond, good looking and a possible embodiment of the Greek god, Adonis. Even though he died in a car crash, his body above the chest was not involved. As I accompanied the funeral director to his embalming area, I noticed an attractive young lady being consoled by a staff member. The funeral director said it was the fiancé of the young airman. He added, "That's' the third one today." Told you he was good looking.

The funeral director and I completed our required examinations and paper work; I thought my responsibility was at an end—not quite. I learned the parents of the young man were flying into town from Greece; they wanted a cremation. Now, Texas had some peculiar laws at the time. Cremations required a casket. The Air Force had peculiar regs as well. The U.S. would pay for either a casket, an urn but not both.

This is where the major, head of finance, I mentioned earlier came into the picture again. I contacted him about the dilemma, and he was adamant that we stick to the regs—casket OR urn—not both. I swear he was still pissed about the time I took the front seat to get flying time.

A lieutenant who was a good friend and who worked in the finance office gave me a hand. This wasn't the first time. Writing a purchase order with funds being allocated by finance first, was a real boo-boo. I don't recall the exact details, but it was a matter of dates on documents—the money was available, but had not been formally released for expenditures. Anyway, he gave me a hand with the dates and that was the end of what could have been a messy situation.

Back to the casket OR urn. The lieutenant and I put our heads together to solve the problem: we had the parents of a dead airman, flying to the U.S. from Greece and we were going to hand them a bill for part of the funeral arrangements. Since the casket was the more expensive of the two items, we opted to have the Air Force pay for it. The cost of the urn was in the neighborhood of $75—I don't remember the details of our little conspiracy, but between the two of us we buried the seventy-five bucks somewhere it would never be noticed. Better that than an international incident over the cost of an urn.

I was the last officer, other than the stay-behind caretaker, to sign off the base. Then it was up to the caretaker and Mother Nature to tend the grass and any critters who entered the base.

13) February 1959 - May 1959, BOQ 6509, Amarillo Air Force Base, Amarillo, Texas

AN AFSC I DIDN'T WANT

Back in the day, each person in the Air Force was identified with an Air Force Specialty Code (AFSC)—Supply Officer was not on my goal list.

The Air Force's next brilliant plan for me—become a Supply Officer. It was a six- to eight-week course, so Pat moved back to Indianapolis for the short stay. Amarillo, located in the Texas panhandle, it's miles and miles of nothing but miles and miles.

On my way into Amarillo, I was westbound on the highway when I noticed a line of trees paralleling the north side of the road. The tops of the trees were bent to the south. I couldn't feel any of the wind affecting the car, so I was confused. I stopped, got out and there wasn't any wind blowing. The prevailing wind in the area was from the north, and that near constant wind caused the trees to grow bent over.

Little of interest about learning the Air Force supply system; how to order, stock and keep track of consumables and fixed property in a unit. Felt like a waste of time when I told them I was going to a Strategic Air Command (SAC) unit and the instructors said SAC had their own system. Oh, well; this too shall pass.

A LITTLE AIR CO0MBAT MANEUVERING (ACM)

[39] North American T-28

Since I needed my four hours of flying time each month to collect flight pay, I was assigned to Base Operations so I could fly back seat in the North American T-28 Trojan. The same birds I saw at Bartow AB, Florida. The ones the later classes were flying. I met with a pilot who was current in the T-28 and would be in the front seat. I saw him talking with another pilot and when he returned, he said the two were going to meet airborne near the base to have simulated combat.

My pilot said we would be over a desolate area, and he would "bounce" the other plane. As planned, the two airplanes met up around 5,000 feet and we found ourselves in a Lufbery circle. Named for Raoul Lufbery, an allied pilot who flew for the Lafayette Escadrille in WW I.

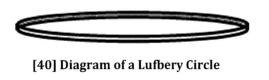

[40] Diagram of a Lufbery Circle

Picture a circle in the horizontal plane with two planes chasing one another around the circumference. The objective is to, in some manner, to fly around the circle faster than the other plane—to get on his tail (6 o'clock position) and shoot him down.

Of course, we didn't have any guns, but if we could get up close to his tail, draw lead on him we could simulate a shoot down by calling, "Guns, guns, guns" over the radio. The problem with this mode of combat is there is no way to break out of the circle; the first one who does is at a disadvantage.

GUNS—GUNS—GUNS

After a couple of turns around the circle, we were losing; my pilot was adding power to go faster and catch up. Never works; the added power, and the attendant added speed caused of the fly a wider circle—old centrifugal force at work. I asked my pilot if I could give it a try. When he said yes, I assumed control of the plane and began easing the power back—speed decreased and we were turning in a smaller circle than our opponent. I also lowered a notch of flaps which allowed me to lower the airspeed and turn tighter. Less than two trips around the circle and we were camped on the other T-28's tail —- guns, guns, guns. My hours of air combat maneuvering flights did teach me a few tricks.

Desolate as this part of Texas was, there was interesting scenery nearby. About twenty-five miles southeast of Amarillo was Palo Duro Canyon State Park. Accessible by car, this gigantic gash in the landscape is around 60 miles long and runs from 800 to 1,000 feet deep. In an area where the land is as flat as a pool table, it was nearly invisible from horse back until quite near to it. This canyon was a fine place for concealment for Native Americans back in the day.

14) May 1959 - July 1959, 104 N.W. 12th Street, Homestead, Florida

This was such a short stay; I don't remember anything about this place except the address. There is a Standard Form we had to keep current for the purposes of obtaining or updating security clearances. It included a section of every residence we occupied. That's where I found this address. Must not have been memorable.

15) July 1959 - February 1960, 2137A Georgia Avenue, Homestead Air Force Base, Florida

PUTTING THAT AFSC TO WORK

On my way from Amarillo, I picked up Pat and our daughter Kathy and headed for Florida. I was assigned to the 19th Bomb Wing of SAC, Wing Supply. I was in charge of the records section which kept track of property owned by the wing. As mentioned in Supply School, this operation looked nothing like what I learned in Amarillo. I set out to OJT (on the job training) myself.

Two weeks after I arrived, we were subjected to an Operational Readiness Inspection (ORI) by the IG (Inspector General). Oh, woe is me. The old master sergeant assigned to inspect my section, wandered around with his clip board asking questions and taking notes. I caught the smell of alcohol any time I was near him. He finished and the IG issued their report. My section was rated Marginal.

When I mentioned to someone the fact the old Sarge was drunk when he was in my area, they said, "You should report him." I answered, "Hell, no. If he'd been sober, my section would have been rated Unsatisfactory!" I figured I could live with Marginal and the only way to go was up.

My supply section improved and the captain I worked for gave me good efficiency reports. For flying time, I assumed I would be flying the T-33 with Base Ops. A friend of mine, and a pilot training classmate, was killed on a night cross country flight in a T-Bird. The Wing Commander decided that all us desk pilots would not fly T-33s with Base Operations. Instead, we were to be assigned to a combat squadron to get our four flying hours per month. Lucky me, I drew a B-47 squadron.

BACK TO FLYING—SORT OF

[41]-Boeing B-47

The bomb wing flew two types of planes and had three squadrons of bombers and one squadron of air refueling planes.

The Boeing B-47 Stratojet was a six-engine jet bomber which came into the SAC fleet in the early 1950s. The three-man crew consisted of a pilot, copilot, and navigator.

The KC-97 carried a crew of five: pilot, copilot, navigator, flight engineer and boom operator. This bird carried four of the largest prop engines ever built — R-4360s. They started with four 7-cylinder radial engines and bolted them together, front to back, to form the 28-cylinder engine on this bird.

[42] Boeing KC-97 Air Refueling Tanker

We lived in Capehart base housing; brand new. So new, the appliance cartons were still sitting in the carport. Again, infinite wisdom convinced the builders to put the housing in a direct path the local blue crab migration. Some decided to detour up and into those boxes, only to learn they couldn't find their way out. You ever smelled a dead crab after a couple of days in the Florida summer sun?

LET'S GO FLYING

Space in a bomber aircraft is at a premium. The pilot and copilot sat in tandem under the canopy on top of the fuselage. The navigator occupied a cramped space a few feet lower than, and in front of the pilots, in the nose of the plane. Those three positions were equipped with ejection seats.

In case of emergencies, the two pilots could eject upward and the navigator's seat ejected downward out of the bottom of the nose. Entry to the B-47 was via a hatch on the lower left side of the fuselage with steps up to a narrow catwalk running on the left side of the pilot's stations. The pilots could climb into their seats from that catwalk; the navigator went down a couple more steps to his position in the nose.

On my first ride, the B-47 Aircraft Commander briefed me on emergencies. In case of bailout, he and the copilot would eject up; the navigator's seat would blow a hatch below his seat and eject him downward. He told me my best chance was to crawl forward and try to fall out the open navigator's hatch. I supposed my next move would be to put my head between my legs and kiss my ass goodbye. I didn't say it—just hoped I wouldn't have to experiment with such an egress. I logged around fifteen hours in the B-47 and once—the AC let me into the copilot's seat and *actually* let me make a turn—a whole great big 30-degree bank turn.

[42a] Unit patch – 19th Air Refueling Squadron

Word came down from upon *high*—the 19th Air Refueling Squadron would pack up, and move *en masse* to Massachusetts—lock, stock and about 18 airplanes. They would need a squadron supply officer—guess who—I drew the short straw again. In actuality, the head of squadron supply was already chosen and I would be second in command.

I stayed at Homestead AFB for a while coordinating the move from the southern end. Just before the full move, I convinced folks my job in Florida was done and I was ready to move. My orders came through, so our family packed up and headed north. Ulterior motive—getting there ahead of the squadron personnel gave us a leg up on the list for base housing.

At Otis AFB, our unit fell under the 8th Air Force which flew B-17s from England during WW II. Also referred to as the Mighty Eighth.

In retirement (Nebraska) I met a gentleman who flew B-17s in WWII and was assigned to the 452nd Heavy Bomb Group – a grandparent unit of the first fighter unit I served with…452nd Fighter Day Squadron.

[42b] Mighty Eighth patch

16) March 1960 - March 1963, 5449C Lemay Avenue, Otis Air Force Base, Massachusetts

Cape Cod was a great place to live. The local restaurants came to recognize us as year-rounders and that paid off in the summer. While a visitor munched on a $12 to $14 lobster, we enjoyed the same meal for around $3. Remember, this was in the 60s.

TANKERS AND MOLE HOLES

This was an Air Defense Command base and our squadron was a tenant unit. Most of the buildings were WW II vintage, as was our Supply Room. They built a new, mostly underground Alert Facility, affectionately known as the Mole Hole. More about this building later.

The basic plane was a modification of the B-29 airframe with a lower deck added. The civilian equivalent was the Boeing B-377. It wasn't in service long due to operating costs—including the fuel it guzzled.

[43] Boeing B-377 Stratocruiser

In the KC-97, the lower deck accommodated the internal fuel tanks of jet fuel for offloading to other airplanes via the boom on the bottom rear of the fuselage. The upper deck also had about seven of these jet-fuel tanks, eight tanks on the lower deck—some were two feet in diameter by twenty feet long. By the time I flew this model, the KC-97G, we were often more limited in range by oil consumption than the fuel available.

[44] Boeing KC-97G

A national crisis, can't remember which, but may have been the Cuban Missile Crisis, put our squadron into a Live-Aboard mode. All aircraft, would be manned with a full crew and actually live aboard their aircraft. During a meeting before the concept went into effect, a plethora of questions came up: How long? How would they be fed?—finally someone asked about bathroom breaks?

The KC-97G only had the most basic fatalities—in a very small "restroom" there was a urinal about a foot in diameter and three feet tall; the sit-down convenience was a pot about two feet in diameter and a foot and a half tall. During normal flight usage, the urinal would be emptied by the ground crew after the flight. As to the pot, the rule was—him who uses; cleans it.

A NOVEL SOLUTION

If the live-aboard lasted more than a day or two, the problem of waste removal could become intolerable. The urinal problem could be handled, but no one seemed to have a solution for the #2 pots. At last, the captain I worked for, got up, took the floor and offered his solution. Roy told the troops that the pots would be lined with large plastic bags to contain the waste. Then he described his idea for disposal. On the ramp, planes were parked wingtip to wingtip and rows of planes were parked nose to nose with enough room to taxi between rows.

Roy continued his explanation. When the plastic bag in the pot was half full, it would be removed from the commode and the neck of the bag twisted and tied with a large twist tie. The bag would be removed from the plane and taken to the center of the taxiway between rows of planes to a point that was equidistant from four planes. A small hole would be cut into the bottom corner of the bag; then the bag would be hoisted overhead and swung around and around until it was empty.

Thinking Roy was offering a serious solution, the room was silent for a few moments. Then the truth caught on, and the room erupted in raucous laughter. The actual solution was a bit simpler—used bags of night soil would be collected by a base services unit.

BACK TO THE SUPPLY ROOM

The military supply system was cumbersome, and next to useless. Example: flight suits—a person needing one should come to the supply room, turn in the worn-out flight suit, then wait until we could requisition one from a depot. In the meantime it was tough luck for the person in need.

We slowly built up a stock of flight suits, flight jackets and other clothing most often in demand. My young three-striper knew his products. Rather than stock six of each, he knew what sizes were in the most demand and ordered accordingly. May seem like a little thing, but he helped us run the supply room smoothly.

AUTO UPDATE

On the Cape were traded the '56 Chevy for a used Buick Roadmaster, 2-door coupe. Pat needed transport as well while I was on Alert duty, so we picked up a Renault Dauphine. Stick shift with four on the floor. Pat didn't care for shifting gears, so I inherited the Renault.

[44a] Renault stick shift

AUTOMATION COMES TO OUR CORNER OF THE WORLD

We got the word from upon high, we would convert our records from hand posted 1120-cards (about 6" x 11") to IBM 80-column punch cards. An IBM punch card machine, similar to the one shown here, was delivered. Fortunately, it came with an operating manual because not a sole in the office knew how to use it. I became responsible for learning the machine, teaching supply office personal how to use it and supervise the transition from the old paper cards to the new and improved IBM cards. Be still my heart.

[45] IBM punch card machine, Model 029

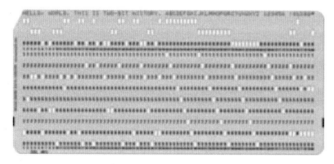

[46] IBM 80 column punch card

The card itself could hold twelve rows of data for each of the 80 columns (left to right). Depending on which hole or holes were punched in a column it was read as a number or letter. Cards were loaded in the hopper, top right, moved downward to the initial point, and then moved from right to left as the keyboard was used to put codes in each of the eighty columns. The info punched into each column was also printed along the top edge of the cards.

In my planning, I estimated the time required to complete the entire transfer within the deadline created by higher headquarters. The task was impossible. I talked to the supply officers in two more tenant squadrons at other bases, and they agreed. We sent the info up the line and our estimates were rejected. All three squadrons began the work. I used my entire complement of airmen in supply and set a schedule.

Since we had one machine, only one person at a time could work on the project. We would work twelve-hour days for Monday through Friday. Saturday would be an eight-hour day and Sunday would be a four-hour shift. We worked that schedule for six to eight weeks until we reached the unrealistic deadline. All three squadrons reported the same results—on deadline day, we all fell short of completing the project. It took a couple more weeks to finish the task.

I believe we were the first of the three squadrons to get the job done, and the other two finished in the next few weeks. I learned another valuable lesson, that higher headquarters is filled with a bunch of idiots who will not listen to the grunts in the field doing the job and who have the best handle on daily life. Later that idea morphed into "despite the staff, the crews will get the job done."

BACK TO THE FLYING LIFE

My squadron commander selected me to move from Supply to become a crew member on the KC-97G tankers. Pat and I took up temporary residence at Randolph AFB, near San Antonio as I completed the combat crew training to be a copilot on a KC-97 combat crew. I remember it was hot and humid that summer, but can't recall much else.

This was the only multi-engine plane I flew that was wide enough the walk around the outside of the flight engineer's station and the copilot's seat to reach my crew position. Most birds like that, forced us to climb over the center console to reach a pilot's seat.

OUR HOST BASE

[46a] Lockheed EC-121C Warning Star

The base host, Air Defense Command operated two distinct and different operations. The primary aircraft assigned to the base was the EC-121C, which was modified from the Lockheed Super Constellation (L-1049)—the one with three tails. Their mission, airborne early warning (AEW), was to look for enemy aircraft approaching the coast in the days before satellites. A few years later, in Korea, these aircraft were involved in a panicky situation which became a humorous incident.

[46b] Early Warning System Radar - a Texas Tower

The second type of early warning radar equipment was installed on multiple Texas Towers. Similar to the offshore oil rigs, these installations contained three huge radomes housing the actual radar dishes. The open deck space allowed helicopters to land so that crews could be shuttled to and from the tower.

One of these towers was lost in extreme weather conditions. The entire structure sank into the ocean along with the full crew aboard. Investigations began, and from the beginning, the fickle finger of fate was being pointed at the captain in command of the tower at the time.

The assertion was that he waited too long before requesting evacuation. It was turning into the old axiom: let's blame the dead guy since he's not here to refute the accusations. Not the first time that tact would be applied, and certainly not the last.

Higher headquarters didn't plan on running into the captain's wife. She refuted their accusations and stated that her husband called her more than once before the tower was lost and told her he had requested evacuation several times and was turned down. I don't recall whose neck they hung the lei of horseshit around, but the captain was not one of them. Another life lesson I kept in mind.

DEUCE

There was a second outfit on this host base. A squadron of the Air Defense Command fighter interceptors flying Deuces. The F-102 was a single seat fighter built by Convair and nicknamed the Delta Dagger due to its delta wings. It's big brother, the F-106, Delta Dart came later and showed up during my assignment in South Korea.

THE MOLE HOLE

When we were on alert, we sent a week at a time in the Mole Hole. The powers preferred Alert Facility, but what else do you call a hole in the ground. In reality, it was a two-story structure with the lower floor below ground, which was dedicated to sleeping and showering. The top floor contained a briefing room, offices and a mess hall. We began pulling alert with two sessions per week; the first was Monday through Thursday and the second stint was Friday through Sunday. So Mondays and Fridays were change-over days for this Three-Four week. After each tour on alert, we had two or three days off.

Apparently, the scheduling was too tough for the staff; balancing the number of days everyone served and keeping it even. So—they presented the problem to the crewmembers—would we like remain on the Three-Four schedule, or—how about seven days in a row and then getting three or four days off. No one was in favor of a whole week in the Hole, so we voted to stay on Three-Four.

The staff had hoped for a different outcome—the staff voted for a week long schedule and their votes counted for more than ours. They got what they wanted, but lost the respect of many of us. An example of the truism I adopted: if you can't afford a No answer, don't ask the question.

PARKING ON A CHRISTMAS TREE

[46c] The Christmas Tree alert parking

The picture, 46c, shows the concrete parking spots for alert aircraft which were located near the Alert Facility. The remains of the Christmas Tree parking area can be seen toward the lower right of the image.

We could accommodate up to seven aircraft on alert at a time. It was a short taxi to the runway shown toward the top left. This is Runway 32.

It was referred to as a Christmas Tree based on its shape. This is a screen shot of a map at Otis AFB base today. The concrete taxiway and the parking spots are crumbling since they have not been used in decades. The Alert Facility was near this spot; while I could not locate any evidence of its existence, I believe it was somewhere along the perimeter to the right or below the Christmas Tree.

A family area was built with chain link fences separating this area from the rest of the Mole Hole secured area. The area had picnic tables and a small playground for kids. Armed guards made it abundantly clear the no one goes in unless authorized. The only time anyone other than staff and crews got in was at an open house held before we formally began alert duty. All the squadron dependents were invited and most attended.

Since the living quarters were below ground, the only light was artificial. When lights went out, it got really dark. There were some hallway lighting low on the walls, but …

My third Aircraft Commander was a weight lifter, and he brought some barbells along on our alert tours. If a practice exercise or the real thing came along, they set off the Klaxon horn—loud, ear piercing and raucous. Jumping into a flight suit in the dark became ordinary, but dodging barbells on the floor—in the dark—could be worth your life. We survived.

ONE STRANGE NIGHT

Due to a fear of a taxi-accident—which could be fatal to a career—the powers generally didn't "exercise the troops" at night. They did however, love to wake us around dawn. When the Klaxon went off, we dressed, got to our plane as fast as possible, started engines and copied a message over the radio from our local Command Post. They were color coded so we could identify what was happening.

The Klaxon went off while it was pitch black outside. This was the first thing that caused the hairs on the back of the neck to perk up. In the airplane, I copied the message. The "color" of the message said it was *not* a practice. More hairs rising. At least it wasn't the color that indicated WW III was imminent—but it was damn serious. We muddled around for a while and finally received the word to shut down engines and return to the Mole Hole.

More muttering and grumbling because we had no idea if "whatever it was" was over or there was more to come. We resigned ourselves to the fact that no one was going to tell us what happened. It was quite a while before we learned the truth.

The BMEWS (Ballistic Missile Early Warning System) was a line of large, land radars that stretched across Canada in those days. They were to look for Russian ballistic missiles heading for the United States. The night we played *let's go to war—almost* —was the night BMEWS went on line live and looking at the horizon. Here's the reason you don't want to buy Version 1.0 of software.

Whoever wrote the program code for the BMEWS radars, forgot to tell them to ignore any signal they acquire that is more that 10 to 15,000 miles away. What comes over the horizon that's about 250,000 miles away? Yep, you got it—the moon. When the moon rose that night, the radars automatically triggered a missile warning. Well, we got a good workout and didn't have to go any farther.

VIP VISITS

[46d] JFK (center, dark suit between 2 AF officers) leaving Air Force One at Otis

There were lighter and more enjoyable times on alert. Remember Hyannis Port and President John F. Kennedy? The Kennedy compound was nearby and Air Force One landed at Otis AFB when the president was coming home to relax. His plane was a VC-137C (tail number 26000) which was a customized Boeing 707—it was similar to the next plane I would fly in my Air Force career.

One afternoon while I was in alert duty, we were told to put on our best bib and tucker.

Well, on alert the only clothing we had were flights suits. We dutifully spruced up what we could, went outside the building and formed up in three rows. When at attention, our heads were supposed to squared sway, straight ahead, keeping eyeballs caged looking forward, no gawking around and remaining silent I was in the front row.

The president was escorted through our compound and passed in front of our formation. My eyes drifted toward him, I guess I caught his attention and he asked something like: How are you? I figured when the Commander in Chief asks a question, I should respond. I maintained enough presence of mind to remember how to address a president. I answered, "Very well, Mr. President." He didn't hang around for anything further and was whisked off by the Secret Service.

One other thing about Air Force One; everybody gets out of the way. One afternoon, I was flying with my crew in the local area on a training mission. A radio call came to us directing us to take our bird about ten miles from the Otis runways and hold until cleared back. Seemed like a strange request until we saw that distinctive blue and white airplane coming in for a landing. We were told we could return to the area in a short while. That's what you call priority traffic.

HOUSING AND WEATHER

Base housing was relatively new compared to the rest of the base. Each building was two stories with a full basement. A group, two or three of these six-unit structures were clustered around two rows of carports. Mostly a close-knit unit, we helped one another, especially when one of the guys was away for temporary duty. This camaraderie held most of the time. One exception—

After a thirty-inch snow, those of us still at home began to dig out the eighteen carports that served three buildings. The one exception lived in the end of our building, and he kept ditzing around his own carport section. I didn't see him move a single shovel-full of snow that wouldn't benefit him. The rest of us plowed through the remaining carports. When we finished, everyone could move their car with ease—except him.

WHAT THE HECK IS REFLEX?

Part of our flying duties involved Reflex to Greenland. Reflex was a fancy word that meant we're going somewhere and stay there for a while—on a regular basis. The flight to Sondrestrom Air Base was about 1,500 nautical miles to the mouth of a fjord half way up the west coast of Greenland. From the coast, we turned

[47] Approaching Runway 09 - Sondrestrom AB, Greenland

northeast and flew about 90 miles to Sondrerstrom Air Base. The runway was built at the bottom of the 1,000 feet deep fjord. We landed on runway 09 (east); takeoffs used the reverse heading 27 (west). There were no go-arounds on landing; put it on the runway or run into the glacier to the east.

Image 47 is a view from a plane approaching the air base. The heading is 09 (east) and this bird is about 300 feet above the ground. The dark bluff to the left is 1,000 feet high. I marked the touchdown end of the runway with a red arrow.

Most of our trips into the base were in daylight and VFR (visual flight rules) conditions. After a week on 24 hour a day alert duty, we would return to Massachusetts. In case WW III kicked off, we would scramble our planes, head to a prearranged location, refuel a nuclear-armed B-52 and return to Sondrestrom. Just to see if we knew what to do, headquarters occasionally "exercised the troops." We raced to our assigned plane, started engines, taxied out, taxied back and parked the plane. Good practice and better than answering to the real thing.

Denmark owns the air base and we were restricted to *our* area (southeast side of the runway, I think). The Base Exchange was on the Danish side of the runway, and we were allowed a trip, once a week and limited to an hour or so.

A DAY OFF—THEY'RE DUMBER THAN A ROCK

On one tour to the remote north, we were allowed to stand down from Alert duty. Don't know why; guess someone decided we wouldn't go to war this day and we could have the day off. My Aircraft Commander (AC), checked out a shotgun, borrowed a jeep from the Motor Pool and invited me to go along with him to hunt Ptarmigans.

If you don't know them, Ptarmigans are an indigenous bird in northern climes. They are about the size of a pigeon, with extra white feathers to ward off the low temps. Did I mention, they are as dumb as a sack of rocks. So, my AC hauled me and his shotgun up the road at the east end of the runway to the top of the fjord. The ground was mostly frozen, but there was green vegetation between the rocky terrain.

We spotted a flock, or herd or whatever of the anticipated quarry. We moved slowly and as quietly as possible and approached the birds. They didn't even look up from their foraging. My AC didn't think it sporting to shoot them on the ground, so we yelled at them a couple of times—head down foraging continued. At last, I picked of several rocks, warned my intrepid shotgunner of my intentions and loosed my missiles at the flock. That did it—the birds took wing and my AC cut loose with his weapon.

He was a decent shot, and one of the wild beasts dropped from the air. We collected the dead bird and returned to base. My AC convinced the mess hall cooks to prepare the bird and he ate it for dinner. I don't recall any of his comments regarding his evening feast, but I've got a feeling it was mainly gamey.

LEAVE THE WIVES AND KIDS BEHIND

Another weather phenomenon; a hurricane running up the east coast was approaching Massachusetts. The powers that be, dictated that the kids and women folk would stay at Otis AFB while us men folk would fly off to more northern climes. I was on a crew assigned to take one of our KC-97G's to Newfoundland, Canada—about 900 miles north. One of the guys on our crew knew the man who was in charge of the officers' club on the base. We put in an order.

Our plane came back to Otis loaded with lobster crates, replete with seaweed and live lobsters. We had a royal picnic in our carport area using every large cooking pot we could lay our hands on.

There were advantages being posted to Cape Cod.

Speaking of kids, our second daughter, Karen, was born in Bourne (a city in Barnstable County; nearby is Buzzards Bay) at the Otis AFB hospital. It was a rambling World War II vintage construction. I remember staying up with Pat while she was in labor, but somewhere after midnight I excused myself for a smoke. I sat on a small divan in a hallway and fell asleep. Later in the morning, a nurse woke me to inform me we had a daughter.

This was the facility where Jackie Kennedy gave birth as well. I'm sure the hospital wing she was in was an upgrade since she was the First Lady. It was a sad time; their son was stillborn.

17) April 1963 - May 1963, BOQ, Castle Air Force Base, Merced, California

[48] Boeing KC-135A Air Refueling Tanker

Never made it to any of the Air Force career enhancing schools for up-and-coming officers. The first such was called Squadron Officers School (SOS), and I was second on the list to attend—PCS (permanent change of station) orders transferring me to a KC-135A Stratotanker refueling squadron in Spokane, Washington, with a detour through Castle AFB in California put the kibosh on me attending SOS.

We took leave and went to Indianapolis to visit our parents. Pat and our children stayed there and I headed to Castle Air Force Base. My dad went with me planning to fly back on his way home from California. The route to the west coast, U.S. 40 was an adventure in its own. In Colorado, the road climbs to 9,000 feet around Rabbit Ears Pass. This is April and the snows covered the mountains. We were inching our way to the top of an incline, about the fifth car in the pack. I could see the lady in the lead car slowing and slowing. I'm yelling, "Don't stop, don't stop." She stopped and there was no way we could gain enough traction to get moving.

A tow truck appeared at the top of the hill. He hooked onto the first car and towed it to the crest. One by one he cleared the stalled cars until he got to me. It was then I learned our good Samaritan was charging $40 for the trip to the top of the hill. I've often wondered if the lady in the lead car was his wife.

U.S. 40, before the day of Interstate highways, was mostly two-lane concrete. At some point we neared the top of another mountain and found a bridge in the middle of an "S" bend. I saw a car approaching from the opposite direction, on the bridge when he lost traction. He was nearly broadside and taking up both lanes. I slowed and eased as far right as I could on a road with next to nothing for shoulders.

Opposite Charlie, horsed his car into his own lane and passed with about two coats of paint to spare. Down to a more reasonable altitude, we stopped at a dinosaur dig. Next, I spotted more snow on the sides of the road. Bad eyesight. This was the Bonneville Salt Flats with all its blowing salt.

A NEW TANKER TO FLY

[48a] KC-135 refueling a B-52

The KC-135 was a variant of the first jet civilian commercial airliner—the Boeing B-707. As an air refueling tanker, the KC-135 had several advantages over the propeller driven KC-97.

It carried more fuel in wing tanks (9,600 gallons) and body tanks (lower deck, centerline 19,500 gallons)—for a total over 29,000 gallons (or nearly 200,000 pounds of fuel).

The entire fuel load was available to offload to jet receivers, as seen in Figure 48a hooked up with a B-52, and all the fuel could be used by the KC-135 itself. Not a good idea to give away *all* the fuel; reserve a bit to get home.

That satisfies the adage; you will always have enough fuel to get to the site of the crash.

With the power available, the KC-135 could fly higher and the increased airspeed made it easier for jet planes to refuel from this airframe. To help with less than desired power on takeoff, we carried water which was injected into the engines to increase thrust. On takeoff, burning the water left four trails of dark black smoke in our wake.

IT WASN'T ALL ROSES

The airplane had a couple of downsides in the design. One was the engines—the original was the Pratt & Whitney J-57 turbojet. It was good in its day, but compared with later power plants, the plane was underpowered. A couple of stories about this shortfall later. It took the Air Force decades to re-engine the plane.

A second problem was the fact that under the wrong circumstances, all that fuel can begin to slosh around in the tanks. It was called Dutch Roll; where the aircraft's nose describes a horizontal figure eight. It has been the reason for the loss of a KC-135 as late as the 2000s. As students, we had to practice Dutch Roll recovery, and this is the closest I ever came to barfing in the air. I was the second student copilot and occupied the navigator's seat while the other student took his turn at Dutch Roll. The navigator's seat is just aft of the pilots and faces the right side of the fuselage. Even though I could rotate the seat to face forward, I was nauseous watching the horizon doing weird things outside the windscreen.

Another drawback was the control system for moving the ailerons, elevators and rudder. The KC-135 did not have any hydraulically boosted control surfaces. Instead, they designed balance tabs which aerodynamically helped move the surfaces. I don't remember the details, but it looked like Rube Goldberg designed the system. It was way better than nothing and is still in use in some airplanes today. In the next section, I'll relate a story about this lack of boosted rudders.

EMERGENCIES ON THE GROUND

In addition to actual flights, we spent hours in a flight simulator that was an exact replica of the pilot end of the plane. This allowed us to practice emergency procedures in a ground environment where a crash was cheaper than dinging a real airplane. These were not the fancy types, suspended on hydraulic legs to simulate movement; these were real big hunks of weight sitting on a concrete floor.

Due to a lucky draw, I was teamed with an experienced tanker pilot. Dan brought a reputation with him as he transitioned from KC-97s to the Stratotanker. Many in the refueling business, knew Big Dan, the Tanker Man.

Back in the day, I smoked—bad enough, but Dan usually had a lit cigar in his mouth. Here's the background when a "Smoke in the cockpit" emergency was simulated. Sitting on top of the instrument shroud at the top of the panel there was a 3" diameter metal pan about a half inch deep with two wires carrying low current electricity pointing down into the pan. A few drops of oil were added to the pan. The instructor/check pilot would throw a switch on his panel and cause the oil in that pan to smoke.

Well, Dan and I were rolling along, handling all the emergencies they threw at us and wondering why no new emergency situations were forthcoming. Finally, the instructor said, "Dan, if you'd put out that damn ceegar, you'd know you got smoke in the cockpit." We ran the checklist, sat back and we all had a good laugh.

Having friends can pay off. Look for Big Dan to make an appearance later on.

18) May 1963 - July 1963, (address unknown), Merced, California

FINISH TRAINING AND HEAD NORTH

I got tired of doing the bachelor bit, and I contacted Pat to make arrangements for her to move to Merced until I graduated. After that, we could drive to Spokane. That worked well and we lived in a small apartment for a couple of months before heading north. I fictionalized that house in one of my thriller novels featuring Alex Hilliard — an air force pilot 😊

My final check ride was not a problem. I passed inspection and became a duly qualified KC-135 copilot. It was time to pack up and drive to Spokane.

CAR SICK

Pat had been in contact with an uncle, her father's brother Dave. He and his wife Sug (short for Sugar) owned a fantastic home in Sausalito, California. They invited us to spend the night with them at the end of our first day on the road. Their home was at the north end of the Golden Gate Bridge with a great view from the huge living room window of San Francisco Bay.

[49] Dave & Sug's fireplace

On that first day traveling north from Merced, one of our daughters decided to be car sick and the other offered sympathetic puking. We left a string of dirty towels in rest area trash cans on our way to San Francisco. So, we arrive at Dave and Sug's home; bedraggled, dirty, smelly and in need of a washing machine. The two could not have been more gracious. We enjoyed a pleasant evening and a restful night. Image 49 is a view from their living room, past the fireplace and toward San Francisco Bay. We left the following morning and headed north again.

Dave was a Navy vet and crewed flying boats. He showed us pictures of his plane, a huge four-engine Mars flying boat, stuck on a sandbar. Pilot error, and they had to wait until high tide and use full engine power to float free and get back to what those behemoths are supposed to do.

SIGHT SEEING

On the way, the scenery was fantastic. One of our stops was at Crater Lake National Park in Oregon. The water is some of the deepest blue you'll ever see and is nearly 2,000 feet deep. It's the seventh largest lake in the world.

[49a] (r to l) Kathy & Karen Snow in July

Crater Lake is about 7,000 feet above sea level and even in summer has snow on the ground. Our daughters got out of the car, raced around the patches of snow, and had a running battle throwing snowballs.

The crater was formed from a volcano and there is an island in the middle. Wizard Island, a mile across, towers 750 feet above the water.

Image 50 shows the retaining wall on a pull-out; Wizard Island is in the center and the far side of the lake can be seen in the background.

[50] Crater Lake and Wizard Island

19) July 1963 - February 1967, 321 Eichenburger Place, Spokane, Washington, Fairchild Air Force Base

GETTING SETTLED

We secured a motel room as temporary quarters as I checked into the base and put our names on the list for base housing. There was Wherry housing within the confines of the base and a cluster of Capehart units about ten miles west of Farichild. The Capehart houses were newer and were available due to the closing of a military facility that was scheduled for the housing.

No brainer—Capehart was newer, removed a bit from the base and in a pine wood area. We moved into the three-bedroom unit and never looked back. I was assigned as a copilot to a combat crew on a KC-135A.

IT GETS COLD IN SPOKANE TOO

[51] Spokane in the winter

The base housing we were assigned was called Geiger Heights. Geiger Field was closed after the housing was built and we became the beneficiaries. It was new; I think we might have been the first occupants. It was around ten miles from Fairchild AFB, but the trip was manageable. The picture [51-Spokane winter] shows our three children playing in the snow.

Our son, Mark was born at the Fairchild AFB, base hospital. Gathered in the street are, Kathy (standing), Karen (in red behind the sled) and Mark (riding the sled).

The image also shows a couple of interesting vehicles. In the foreground is a black four-door Simca circa mid-50s model, probably the Aronde model. It was French built and I got it from the only Simca mechanic in the area. It was used, well used. I replaced the inside door panels and installed seat belts for both front seats. I drove it to work so Pat could have the Olds at home. I re-learned how to drive a stick shift—again. It was reasonable dependable and I can remember only one time she didn't start after sitting outside for a seven-day alert tour. A short push by a friend and I was chugging along as fast as the little four-cylinder engine could manage.

In the back ground is a white "SUV" of its day. Don't remember the make or model, but it belonged to a good neighbor and friend. Their daughter baby-sat our kids on many occasions. Later, at survival school I was without transportation, and I lost the heel of a boot when an idiot stepped on the back of my foot. I desperately needed a new pair, and this neighbor, John Humphrey, loaned me his vehicle so I could do some shopping. I found new boots and survived the rest of the survival course. Great neighbor. Why do the good ones die young? I remember hearing a couple of years later that cancer took John.

IT HAPPENED BEFORE I ARRIVED

There was a unit, the 1st Combat Evaluation Group (CEG) based at Barksdale AFB, Louisiana. They visited each base to tell crew members how badly they were doing. One of the CEG Aircraft Commanders was Major Ralph W. Trousdale—call me Trous, he would add to any introduction. Trous had forgotten more about the KC-135 than most of us knew.

A few months before my arrival at Fairchild AFB, Trous visited the base and ripped a few knickers in the process. I arrived, and the fickle finger pointed to Trous and he was transferred to Fairchild as well. Those folks with the ripped knickers were lying in wait for him. Trous was scheduled for a check ride before he was assigned a crew and I would fly copilot with him. The chief standboard crew was to administer the check ride. Standboard crews are the local equivalent of the CEG.

NOW I'M IN THE MIDDLE

Needless to say, Trous did not pass his check ride. The chief standboard pilot went out of his way to nitpick everything Trous did. I made out okay since it was not a graded flight for me. Trous and I were paired on a crew to practice and prepare for a re-check. We both passed the next ride and went about becoming a cohesive crew.

As luck would have it, CEG came to town a month later. They busted i.e., failed, all the local standboard pilots. The squadron commander looked around for replacements for senior standboard pilots. He picked Trous, and I became the standboard copilot on the senior crew. I could not have found a better mentor than Trous if I advertised for one.

As we finished a training mission, I made the landing and we were rolling down the runway. We typically let the bird slow down on its own before we braked for the turnoff at the end of the runway. Not a good practice if another plane was landing close behind us, but today all was clear. Trous asked if I'd ever bicycled the 135. He took control of the plane and rolled the control wheel right and left and let the ailerons turn the plane aerodynamically. He gave control back to me and I added the same control input for a few turns back and forth. The turns were gentle but they definitely allowed a change in direction here on the ground for small turns. Look for this technique to pop up later.

[51a] Unit patch – 92nd Air Refueling Squadron, Fairchild AFB

Trous was more than a mentor. In a way, he was almost a father figure. Look for him later and check out two short stories, Grace Period I & Grace Period II in the Appendix.

ENOUGH TIME IN THE RIGHT SEAT

[52] Author in right seat

For the life of me, I cannot remember the reason I flew with Trous to the South Pacific. I must have because I have pics from the time and I remember landing at Andersen Air Force Base on the island of Guam. The Guam runway is a swayback affair and at the end of the runway is a tall cliff. Clearing the cliff edge on takeoff, we immediately had 500 feet of altitude.

Shortly after departure, we flew over a pair of islands shown in Figure 53 peeking out from under afternoon cumulous build-ups. The names are Saipan and Tinian; the B-29 Enola Gay launched from Tinian to drop the atomic bomb on Japan. I think the island bottom right (Figure 53) is Tinian.

[53] Saipan and Tinian

Trous nominated me for upgrade to Aircraft Commander. Our squadron commander, Lt. Colonel Smith, scheduled me for a five-hour flight so he could decide if I was ready. In a side by side multi-engine plane, the copilot controls the bird with a right hand on the control wheel and the left hand on the throttles. Moving to the left seat meant all that was reversed. I found I had enough strength in my left arm, but five solid hours of flying the local area was a real strain. We completed a wide range of maneuvers and a number of touch-and-go landings—sometimes referred to as a Crash and Dash.

I was totally wrung out from the flight, but I passed muster in my squadron commander's eyes. I got a formal check ride and assumed command of my own crew. Flying in SAC (Strategic Air Command), we were subject to strict rules. One was takeoff time. Each flight was scheduled to take off at a specific time; the takeoff roll was to start at that time—not a second earlier—and no later than five minutes past the scheduled time.

FIRST TIME OUT

My first flight with my crew and on my own—of course there was an Instructor Pilot riding the jump seat—just in case—became a near panic situation from the get-go. Maintenance was still working on the plane when we arrived to pre-flight the bird. We worked around them as best we could and they finally released the plane to us about fifteen minutes before our takeoff time. We cranked engines, taxied toward the runway and ran all the checklists. A call came from the navigator: approaching takeoff time.

We were on a taxiway parallel to the runway going the opposite direction from takeoff. I pressed ahead as fast as we could go within the limits of safety. We were about fifty yards from the first of two ninety-degree turns which would put us on the runway for takeoff. My Nav called again: thirty seconds to takeoff plus five.

I took a deep breath and keyed the radio button: Fairchild Tower, this is (callsign) rolling.

The tower responded: Roger, (callsign) I have you on the takeoff roll.

I used minimum braking to stay safe, made the two turns using nose wheel steering and differential power on the engines, and lined up on the runway heading. Without stopping, I eased the throttles up to full power and off we went. Probably the only pilot who logged an on-time takeoff with two ninety-degree turns to go, but it was a counter in the plus column and not a "not on time" takeoff on my first flight as AC. Airborne, the instructor pilot clapped me on the shoulder and said: good job.

GO FOR SOMETHING DIFFERENT

Each member of my crew grew stronger at their flying duties. I knew I could trust them to do their job and help keep us all out of trouble. I checked with my squadron operations officer about sending my crew to Eielson AFB, Alaska to participate in the Eielson Tanker Task Force (ETTF). The mission supported the reconnaissance operations toward the Russian coast to the west and a mission flying east to Greenland to refuel a B-52 on airborne alert.

PAYS TO BE PUSHY

The Ops officer declined saying they were concerned about sending young navigators on such a challenging mission. Basically, you can't go until you've been there—kind of a Catch 22. My Nav was young, but he was good. A week or so later, he cornered the wing commander at a Saturday night dance at the Officers' Club and pleaded his case. Not usually a good idea to go out of channels like that, but he won the day with the commander. We were on the schedule for the Eielson TTF.

IT CAN GET COLD IN ALASKA

We flew some missions heading west from the center of Alaska, and refueled RC-135D's who spent hours patrolling the Russian coast from the Bering Sea off northeast Russia to the Barents Sea off the far northwest coast. The other mission involved a flight from Eielson AFB over the pole to Thule, Greenland. There we rendezvoused with a nuclear armed B-52 flying a 24-hour mission code named Chrome Dome. We offloaded fuel to the bomber and headed back home.

My navigator was a busy boy. We tracked outbound on a radial of the Thule TACAN (tactical air navigation), a radio station that provides radial and distance information. We tracked the TACAN for 105 nautical miles. That gave the Nav a fixed radial and distance point to help set his course back to Alaska some 2,000 miles away. There was nothing below us except snow and ice for several hours. No navigation aids, no landmarks; only the skill of my navigator got us home. As I recall, we hit the coast of Alaska within twenty miles of the intended point. Damn good work, Nav.

The entire family, ca 1964/65. L to R: Pat, Kathy, Author, Mark and Karen.

Used a tripod with the camera on timer to get this shot.

[53a] Full family Spokane

BEEN NORTH, LET'S GO SOUTH

The next challenge for my crew was to get in line for a Young Tiger tour to South East Asia (SEA). My first problem was a requirement the AC be passenger qualified, which was based on total flying hours. To become pax qualified I would need to accumulate 2,000 hours in the air—1,999 hours, no go; 2,001 hours and you became a Senior Pilot and qualified to haul passengers. Must be a reason behind it, but it's not obvious.

I was about 100 hours short, so I began a campaign to build my hours. I hooked a pair of rides on B-52 Chrome Dome flights. These bombers were flying airborne alert flights of 24 hours each and were armed with nuclear weapons. Two of those flights and I logged 40 to 50 hours. A few extra flights, local and cross-country, left me about 15 hours short of the goal.

My squadron commander went to bat for me, saying he'll be passenger qualified by the time he gets there. The upstream powers relented and my crew was scheduled for a Young Tiger ninety-day tour to SEA. Another Aircraft Commander was sent to SEA with less that the required 2,000 hours. Unfortunately, he was involved in a screw-up, crashed and some died in the accident. I was the last crew commander who was allowed to do the SEA tour without the requisite number of hours.

On one leg of our trip to the Far East, we carried quite a load. My Boom Operator oversaw the loading of 70 passengers, 5,000 pounds of miscellaneous cargo and a J-57 jet engine. Fortunately, the engine was mounted on a mobile, cart with wheels. Did that make us a five-engine jet plane? No, but as a Senior Pilot, I was passenger qualified.

WE'RE HERE—NOW WHAT

The SEA tour would include time at Kadena AB, Okinawa (now Japan) with two or three, one-week stints in Thailand. In Thailand, we flew missions from U-tapao, the civilian airport in Bangkok and Takhli Royal Thai Air Force Base in central Thailand.

HANDLING THE BIG GUYS

We flew B-52 refueling missions from Kadena, which usually was 18 KC-135s in the stream. These flights were in support of Arc Light Operations. The bombers, from Andersen AFB, Guam, were headed for strategic targets in South Vietnam, and North Vietnam. Figure 54 is a stylized view of our trip to refuel the bombers. We flew in-trail from Kadena [1] until we neared the refueling area [2]. See Figure 54a for details of the separation.

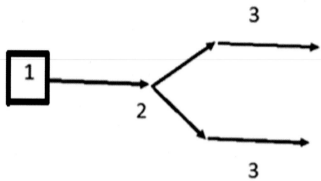

[54] Overview of a bomber stream refueling

At point [2] the tanker stream separated left and right and then turned forty-five degrees to set up the two refueling tracks [3]. Somewhere before point [2], the tankers moved into three ship elements. In each three-ship element, number Two and number Three were about thirty-degrees off (behind) and a half mile to the right and five-hundred feet above the ship in front of us. At point [2], the three-ship elements would alternate turning right or left.

Enroute to the refueling area we flew in-trail formation (Figure 54a) using one-mile horizontal and 500 feet vertical separation.

Just prior to Point 2

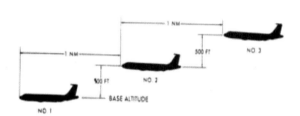

[54a] Enroute tanker formation

(Figure 54b), we moved into refueling formation (Figure 54b). For bomber refueling the horizontal separation was two miles at a 60-degree angle on lead. Again, we were stacked up 500 feet for vertical separation.

When refueling fighters, the horizontal separation was reduced to one mile.

Before I continue, a few words about formation flying. Here (Figure 54b) is a rough picture of the three-ship elements we were flying.

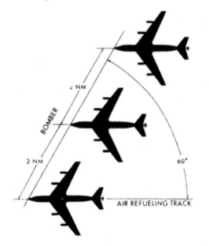

[54b] Tanker formation for bomber stream refueling

RULES OF THE ROAD

A rule of formation flying is to avoid turning into the echelon. In the above case, a right echelon formation; if lead turns left, he is turning away from the echelon. The opposite is true if lead turns right. This is the reason why: in our case, if Lead turns right, Number Two has to fly a smaller circle in the turn, requiring him to reduce power. The same is true for Three who is flying an even smaller circle. With only 1,000 feet difference in altitude between Lead and Three, the complications were not insurmountable. Now for the night in question.

I TOLD YOU NIGHT FLYING IS SCARY

I was assigned the position of, Primary Ground Spare—if any of the regulars aborted before takeoff, we would go as a fill-in tanker. This night at Kadena was miserable, pitch black with blowing rain. We got our plane to the end of the runway, parked on the large pad which was out of the way of other planes, and watched as the tankers took off one by one. There was a lot of chatter on the radio as the planes staggered into the air.

One genius in a staff car, called Mobile Control, commented on how long it was taking the tankers to get airborne. After he made the comment a second time, I keyed my radio and said, "Take a look at the tailwind component." With a wind on your tail, it takes more runway to get off the ground. No more chatter about long takeoff rolls, but they weren't likely to switch to the other runway.

We recalculated takeoff performance and we were safe as long as the winds didn't get any worse. One by one, the other tankers taxied by us and took off. Then the call came, one plane said he was aborting on the ground. Oh, goody—here we go—it was time to hurl our bodies into that pitch-black night. We moved into line to fill the empty slot by the one who aborted.

AN UNUSUAL SITUATION—A 4-SHIP FORMATION

Only one drawback for us; to make the mission work out, I had to fill in as Number Four in this element.

That put me 1,500 feet above Lead's altitude. In these hot climates, that extra 500 feet would make a great deal of difference in performance. As we approached the split point (2 in Figure 54), I knew Lead would make a right turn—into the echelon. I did my best to prepare for the turn, but it wasn't quite enough. I was using full military power on all engines and tried to maintain altitude—it was a losing cause. I was drifting backward from the others and sacrificed a bit of altitude to gain airspeed.

At rollout on the refueling track, I was behind the position I should be in, but with full power still on I was gaining. I was back in formation by the time the bombers put in an appearance. Sweaty palms, but all was well.

WHAT THE HELL IS JUMPING JACK?

Never ignore the obvious. During pre-flight briefing, Jumping Jack was mentioned. I asked our Lead pilot what Jumping Jack was. He said: Don't sweat it, we never get tasked for it. That was the obvious clue—after refueling the bombers, Lead called over the radio—we got Jumping Jack. Before I could say anything, Two called "low oil pressure on number three." I knew Lead wouldn't volunteer and that left me as the goat. My crew scrambled through our paper work and found the details. We headed for a specified point and stayed there for a while in case any planes needed radio relay of messages. No one did, so we headed back to Kadena a couple of hours later than usual.

At debriefing and after the paperwork was done, they repeated an old WW II practice of giving a combat ration—of whiskey—to the returning crew members. We drank it, but it was mainly rot-gut and I wondered if it was left over from WW II.

At the end of one mission, I decided to demonstrate an automatic approach to landing. Our autopilot was good, but nowhere near the auto-landing types they have today. It would lock onto an ILS (Instrument Landing System) radio, line us up on final approach and descend down the glide slope. At minimums, the pilot would take over and make the landing.

My copilot set up the autopilot, engaged it and was monitoring the plane's instruments as we came in for the landing. The skies were clear of clouds and it was a beautiful morning. My view out the windscreen didn't seem right. Then my navigator called on the intercom and said we were a mile left of centerline. He had been monitoring our approach on his radar equipment. I called the tower, reported our situation, disengaged the autopilot and the copilot completed the landing without incident.

We debriefed the A&E (Avionics & Electronics) techs. ILS ground systems put out a correct identifier and we received it loud and clear. The ILS instrument in the cockpit had a red OFF flag for both course and glide slope—neither flag came up. No one had an answer why the electronics messed up with no warnings to the crew. Another lesson learned; check, check and double check—anticipate the worst and stay alive.

OUTSIDE THE CHRONOLOGY

I haven't introduced these planes yet, but it seems appropriate now. Rivet Amber, was an RC-135E and carried what must have been the world's largest airborne radar. I wasn't qualified to fly that one, but I flew the RC-135S, Rivet Ball. The loss of Rivet Ball which I was flying happened on 13 January 1969 and it is covered in detail in a section titled: THE OTHER HALF OF OUR EIELSON DUTIES and in my memoir "The Iron Pumpkin." That date was five months before this trip to Okinawa

[54c] Rivet Amber

Rivet Amber's radome is visible in Figure 54c, the darker area wrapping around the fuselage just forward of the right wing root.

She was lost on 5 June 1969 on a ferry flight from Shemya AS to Eielson AFB. Rumors abound, but the only thing we know for certain is that *Rivet Amber* and crew were *lost* over the Bering Sea

With 19 Souls On Board, we all lost good friends and squadron mates. One of the maintenance men onboard my plane when we crashed had been transferred to Amber and was lost as well.

One of the most eloquent remembrances appeared in the Fairbanks Daily News Minor newspaper on June 21, 1969.

Dear Editor:

On the recent disappearance of the KC 135 plane, one might read from the news prints and wonder out of curiosity who they were or what kind of people they were. Maybe I can answer both of these questions.

When I started working on the military bases years ago, I tried not to get deeply involved with military families, because they would be stationed here one day and rotate the next. But in spite of yourself you do become involved with their troubles and triumphs, ups and downs, sports, politics and whatever and their kids look to you from time to time for advice. It's the kind of involvement that makes you feel good to be a part of their lives, so if the good Lord called his children home; He called nineteen of the greatest guys I've known. Nineteen men doing their "thing" to help keep this the greatest nation on earth.

So, if anyone should wonder who they were or what kind of people they were, well, they were the greatest, they were my friends, and I knew them well.

Sincerely,

Jay Spearman

(their barber)

Thank you, Jay. You were my barber too.

… and with a sad heart, I will return to my narrative.

REFUELING OUR LITTLE FRIENDS

The other half of the Young Tiger tour was flying from Thailand refueling fighter aircraft fighting the air war in Vietnam. We refueled a few F-104s, gobs of F-4s whose job was to protect against MiGs coming from way north, and F-105s (Thuds) loaded to the max for their bombing missions in Route Pack 6; the Hanoi area. In Figure 55, I'm standing on the steps of our "hooch" at Takhli. It was raised above ground for two reasons: high water and snakes. Snakes came in two versions: a one-stepper and a three-stepper. That's the distance you could expect to run if bitten.

[55]-John's "Hooch" at Takhli RTAFB

Note the colors, tan "sidewalk" and brown building. Everything else was green; foliage, trees, grass and fungi.

My boom operator, the man in charge of refueling anything that came up behind us, was a young three-striper. In those days that rank was an Airman First Class. Most Boomers were Tech and Master Sergeants. He was good, and those were not my words. Compliments came from those fighter pilots who knew that in minutes from the time they dropped off the boom, they would be facing life and death situations.

The refueling receptacle on the F-4 Phantom was centerline aft of the large cockpit canopy which covered a pair of tandem cockpits. The fighter stabilized in refueling position and the pilot told my Boomer his refueling receptacle door would not open; it was stuck. My Boomer directed the fighter to move a bit closer, reached over the heads of the pilot and his GIB (Guy in Back) with the tip of his boom and tapped on the recalcitrant door—it popped open and we offloaded fuel to a grateful F-4 crew. Only one of the times those fighter jocks sent a compliment his way.

ASIDE

Back at Fairchild, after this SEA tour, my boom operator failed a written test about who knows what. This could be a career buster. I went to my squadron commander and discussed the situation. I told him about the job Boomer did in the air over Vietnam, and told him I hoped he did not expect to see this test failure noted in an Effectiveness Report that was due shortly. He agreed.

PASS THE GAS

Typically, we had more fuel that we needed when we finished a mission. The first time my Boomer said, "he has the briefed offload, should I disconnect?" — my answer was—leave him on the boom until you get a pressure disconnect. When the fighter was full, and back pressure occurred, the boom system would make the disconnect on its own. I think we had a lot of satisfied customers.

A typical offload to a fighter was 1.8k—1.800 pounds of fuel (around 250 gallons). One, by the book, tanker pilot would tell his boom operator to toggle a disconnect as soon as the fighter had the prescribed amount of fuel. He didn't last long. The tankers had plenty of fuel; no sense in being stingy.

I remember an evening at the officers' club at Takhli (Royal Thai Air Force Base – central Thailand) we were surrounded by F-105 pilots. Their comments were aimed at us—crews from the that big, lumbering airplane. I caught the attention of a Thud pilot sitting nearby, waggled an index finger at him and said, "That's the most important finger in the world to you." He looked perplexed, and I added, "That's the finger that turns on the refueling pumps when you're hanging on my boom." He roared with laughter, as did his friends, and we didn't have to buy our drinks for the rest of that evening.

On one of these flights, we learned another lesson. We normally left 5,000 pounds of fuel in the forward body tank to give us a solid center of gravity for landing. This night my copilot and I were distracted and we were down to 3,000 pounds in that tank.

Center of gravity was within limits so we completed our landing. As I slowed on the runway and turned onto a taxiway, the bird felt light on the nose gear. It seemed as if it would like to lift up from the concrete. I kept extra power on the engines which tended to put more force on the nose wheel. I didn't reduce that extra power until we passed our tail stand out a hatch and the ground crew installed it under the tail. I doubt the bird would have tipped backward on her tail, but I wasn't about to find out. There aren't many excuses for allowing an airplane to do that. Bad kind of a taxi accident avoided.

GOING NORTH

The Mekong River separates Thailand and Laos; we were directed to remain on the Thai side, south of the river. We had refueling orbits all along the Thai border, and did our best not to violate Laotian airspace. Occasionally, we would let a wingtip drag over the border—that was the difference whether our time in the air was designated as Combat time or Combat Support time.

At times, we would be assigned to one of the border orbits without any specific receivers. We could offload fuel to anyone coming south from strikes in the north and who needed fuel. I heard a call from an F-105 coming out of the Thud Ridge area near Hanoi; he had major battle damage and needed fuel. I completed our turn and headed north while bringing the throttles up to max power. A GCI (Ground Control Intercept) station was talking to the fighter and I came up on his frequency. I told the GCI controller I would go after the fighter.

The controller gave us vectors to us so we could intercept the damaged plane. I sent my Boomer back to the boom pod so we could refuel the F-105 as soon as the rendezvous was complete. Boomer asked if he should extend the boom to be ready. I gave him an emphatic NO. We were well above the boom lowering airspeed limit—putting it down now would likely tear the damn thing off the plane. I did tell him to keep an eye on the ground below us looking for pink puffs. He asked what that would be—he was quiet when I told him that would be anti-aircraft guns firing at us and if you could see pink, the aim was pretty damn close to us.

TAKEOVER

The ground controller kept giving us headings, but it was taking too long. We should have been near him by now. Our primary communications radio was the UHF (Ultra High Frequency), which we all were using had an extra feature—UHF – DF (Direction Finding). I asked the fighter to give us a short count – one to five and back again so the DF needle in our cockpit would point at him. Damn.

The DF needle pointed at our tail—crap. The damaged fighter was south of us while we were still northbound. I hauled the bird around in the tightest turn I could manage and planned to catch the fighter from behind. The GCI controller called off the chase and sent us home. I later learned that the F-105 went down, but at least he was south of the Mekong River and was in friendly territory. The bird was lost, but the pilot survived.

I kicked myself in the butt for not thinking of the DF radio function earlier in the drama. Maybe we could have saved the plane. My major mistake was thinking that the ground controller was painting the fighter on his radar. In hind sight, I should have realized his radar didn't reach far enough to identify the fighter and he was using his best judgement. What if? What if? What if? I had to be satisfied with the fact the F-105 pilot was safe.

WHERE WERE WE?

The next problem was what to tell the powers about our excursion into northern Laos. Far beyond any dipping a wingtip over the river. My navigator asked me if he should show where we had been on his map and nav log. I told him and the rest of the crew; be honest, don't lie, if questioned tell them you let me know how far we had exceeded our area of operations, that I acknowledged your information and kept going on my own. My theory was if we were heroes, I'd share the glory; if we were goats, I would take the lumps.

FACTS OF LIFE

I think that attitude rubbed off on them. I also made those who worked for me write their own Effectiveness Reports. The first time I wrote one on a subordinate, I was completely lost and in the dark. Luck was with the guy working for me. My boss had a talk with me and I learned the facts of life. ER's were always inflated; to do less was to sign a career death warrant for the person.

I told them why I was forcing them to do the writing—based on my first time out. Also, they would likely bring up points I might have missed or forgotten. On top of that, I assured them I would rewrite the ER myself to be sure they got the best shake and boost to their career. My luck—I was blessed to have people working for me who deserved to move up in rank. I can't say the same of a few of the whack-a-doodles I worked for.

TELL THE TRUTH

My Nav turned in all his paperwork at the end of the flight as usual. The information plotted on his map and in his log was totally accurate. Funny, nobody ever mentioned a thing about our excursion into Laos.

WE DID HAVE SOME FUN OCCASIONALLY

[55a]- (l-r) Larry, Author, Thai family

A young Thai man (Figure 55a, right) introduced himself to us one day saying his company would send him to work in the U.S. if his English was good enough. He wanted to practice on us.

My copilot and I are on the left in Figure 55 and his children are front and center. He invited us to his home in the morning for breakfast. Knowing the American stomachs are sensitive to many foreign cuisine, he offered us bottled orange cola and bananas.

Following this repast, he invited us to accompany him and feed the Monks.

[55b] (r to l) Host, my copilot, and a Monk

We felt a bit awkward, but our guest provided us with the food we would share. The Monks make the rounds each morning and collect whatever offerings are available. The amount collected must sustain them for the entire day.

My copilot, Figure 55b is sharing food with the Monk in his saffron colored robe.

This next one surprised even me. Years later, I discovered the image of me riding an elephant. I'd totally forgotten this wild ride. Not sure exactly where or when I climbed upon the back of this beast.

As you can see in Figure 55c, this was a baby elephant; I'm guessing about five feet tall. It was also tethered to the tree on the right.

If you can see my face, I'm looking apprehensive. I survived my ride with nary an injury and one of my crewmembers took the picture.

55c Yee Haw! Buckeroo

A GOING AWAY PRESENT

Our temporary duty time was up, so we packed up to go home. Got a feeling that our local maids kept track of the length of our tour and let the "stealy-boys" know this was our last night in the BOQ. Something woke me in the middle of the night, but I rolled over and went back to sleep. Good thing, I was told the thieves often wore razor blades taped to the back of their hands so one swipe could inflict serious damage.

When we woke in the morning, we saw someone had rummaged around the room. With all the stuff we were hauling home, they only took cash. My wallet was missing, fortunately I found it in the bushes outside our room. Cash was gone, but everything else was still there. They were smart; robbing us on the last night of our tour meant we couldn't hang around to help in the criminal investigation. We made a quick report to the Air Police and cranked up our bird and headed home.

I think I was the youngest aircraft commander to take a KC-135 tanker to participate in the air war in Southeast Asia. I'm proud of the contributions my crew accomplished. End of my SEA tour in tankers – but I'll be back.

We flew non-stop back the March AFB, California where we refueled the plane. We were near the takeoff end of the runway, when the tower called us. There was an airman at Base Ops who was on emergency leave and needed to get back to the Spokane area. I parked on the runup pad at the end of the runway. The tower said they would handle the paperwork and bring him to our plane in a staff car.

I left the engines running and we opened the front entrance hatch. Our Boomer climbed down the ladder and helped the airman into the plane. It was quite a chore because the airman on leave was using a pair of crutches. We got him home.

BACK AT FAIRCHILD – MORE FAILED CHECK RIDES

Both pilots on another crew failed the annual checkride and the staff looked for a solution to lend a hand to all involved. I've mentioned, that my copilot was a strong crew member, and contributed to a solution. Put my copilot up with the pilot who failed his ride, and pass the failed copilot to me for remedial work.

The change was completed, and our crew pulled a seven-day tour alert duty. We flight planned for a training mission coming off alert. My instructions to my new copilot were; for the first couple of flights, you do copilot things and I'll do the pilot things. After that, we'll press on as needed. He seemed to be okay with this approach.

He flew two training flights in the right seat and was an accomplished crew member. I had no real complaints. On our third flight, I told him, "This is your mission; you do all the pilot things still flying from the right seat." He seemed surprised, but the mission went off without a hitch. As our B-52 approached from behind us for the air refueling, he asked whether it was okay for him to be in charge. I told him, I'm not sure exactly what the protocols were, but we won't bother asking the question because we don't want a no answer.

Again, he did a creditable job flying and completing the mission. I had been digging into his background; he came out of pilot flight training having flown, the T-38, Talon. The Talon was a light, nimble twin-engine jet trainer capable of exceeding Mach 1. Now he was herding hundreds of thousand pounds of airplane in the form of a four-engine KC-135 around the skies. No wonder his aircraft skills were weak. I assigned him a batch of time in the base flight simulator to practice flight maneuvers in visual and instrument conditions.

With a dozen hours or so in the simulator, his flying skills were up to that of any pilot in the squadron. On his last crew, he was allowed to vegetate in the right seat. Given the opportunity to expand his skills, he became an outstanding crew member. All he needed was the opportunity to stretch his wings and soar with the eagles.

20) February - March 1967, Castle Air Force Base, Merced Air Force Base, California

TIME TO MOVE AGAIN

My next PCS orders arrived. I would be going to the 24th Strategic Reconnaissance Squadron at Eielson AFB, Alaska. On the way, the Air Force graciously allowed me to attend survival school. Due to the reconnaissance assignment, this was an advanced course relating to my upcoming duties.

THIS SOUNDS LIKE FUN, NOT ...

The school, run by a special unit stationed at Fairchild AFB, allowed us to crawl through two or three miles—more like a hundred yards—of humps, hillocks, holes on an obstacle course until we reached the barbed wire. It made no matter how well we navigated the course; the end was inevitable; you would be captured and held as a prisoner.

I crawled, and crawled—then I spotted a hole, maybe eight to ten feet long, a couple of feet across and five feet deep. Looked like a good place to rest, take a look at the terrain and plan my next move. I hit the bottom near one end, ducked down and looked toward the other end of my sanctuary and saw a pair of instructors sneaking a smoke. Now this might be a problem; instructors had the power to tag students and make them repeat the course. I also figured they were violating their orders—I didn't acknowledge them, just climbed out of our trench and crawled away. I didn't mention this episode to anyone and apparently, neither did they.

We were interrogated, marched single file with hoods over our heads and locked into our cells built into an old abandoned ammunition bunker. The most memorable character there was Master Sargent Sepp.

For more about this man, see "Awards and Other Stuff" in the Appendix B of this book. Another man who caught my eye was a stocky gent with rusty hair and a walrus mustache. He seemed to be more of an observer than a staff member. Later, I asked and got the answer. He was a CPO (Chief Petty Officer) stationed aboard the U.S.S. Pueblo; the Navy reconnaissance ship captured by the North Koreans. This CPO is also the one who suffered severe beatings. He was observing this operation so he could set up a similar school for the Navy.

THIS TREK AIN'T NO PICNIC

After the confinement, we were graciously allowed to go on a trek in the woods in Washington near the Idaho—Washington state line north of Spokane. We were divided into pairs to make the three-day march. It felt like ten miles, but later I measured it on the map they gave us and it was barely two and a half miles. Here's why...

The area sustained a heavy forest fire a few years back and many large trees were left along with dense secondary growth of bushes and saplings. We carried our meager belongings rolled up in a shelter half tied with parachute cords. The temps were in the forties when the sun was up and dropped to near freezing at night.

Since I was the ranking officer of our group of seven or eight, I was named the leader. I was also probably the least knowledgeable person about camping out. On our first day out, one of the NCOs in the group scrounged around until he located a sturdy branch and whittled it down to the shape of a "Y." We stopped of a rest break and were sitting around a small circle of undergrowth when this NCO produced a length of surgical tubing from his pocket. He attached it to his Y-branch and voila – he owned a slingshot. He had also located a few small, round rocks and was hefting it and looked around for a target.

A rustling in the vegetation nearby and the Sarge spotted a rabbit trying to sneak away. One rock, one shot and that rabbit was dead. Our instructor used this as a learning session. He demonstrated how to skin an animal—even though the rabbit was small, the techniques would apply larger kills. There wasn't enough meat on this hare to split, so he went into a stew of potatoes and onions which had been issued to each group.

Before we got to supper, our instructor also said we shouldn't waste any part of this find. He talked about using the pelt and then he came to the head. He popped an eyeball out of that dead rabbit, held it in his palm and clapped the hand over his mouth. He swallowed while telling us how nutritious it was. He looked around the group and asked who wanted the other eye. There were no volunteers and being the "leader," I said yes. I had to swallow that eyeball at least three times before it stayed down. We enjoyed our rabbit stew while the other groups had only potatoes and onions. Another notch on the axiom that good NCOs run the military.

DAY TWO OF TREKKING

Due to the low temps, tying our packs could be dangerous; too much pressure on the inside of finger joints, could rub the skin raw. Open sores were risky, since we had no medical supplies. Gloves were the only preventative action available. That pack on our backs was not well balanced and caused all manner of unanticipated problems.

At times the secondary growth of one-to-two-inch thick saplings was so dense we had to push them apart to make headway. That also meant twisting sideways to wedge ourselves and the pack through the opening. That pack caused another problem for me.

When we were lucky enough to find a large fallen tree left from the fire and it paralleled our compass heading, we would climb up on it and walk the tree, free from the undergrowth. On one such occasion, I was moving behind my partner, my foot slipped and I fell off our log. As I went down, I twisted my body so I would land on my back, rather land on the backpack. There it was, not quite on the ground, being supported by undergrowth and knowing what a turtle on his back feels like.

That image came to my mind and I was laughing out loud. My partner thought I was in pain and nearly panicked. I assured him all was well and with a bit of help, the turtle remounted the log and we pressed on. We maintained a continuous vigil for the instructors bopping through the area wearing red knit caps. I suppose the bright colored caps were to give us the advantage of spotting them and hide.

My partner and I came upon a long downhill section of our proposed route. Must have been a hundred yards to the tree line below and we pressed forward. About half way down was a scrawny tree, maybe six feet tall. We were pooped and decided to sit down by the tree and take five. As we stood up to leave we were adjusting our backpacks; I was facing uphill. A *red hat* came out of the tree line at the top of the hill. I looked at my partner and said, "Make like a tree and don't move a muscle."

The instructor in a red stocking cap came bounding down the hill and stopped next to our tree. He said, "What compass course are you following?" We gave him the course we were on and he said, "That's a good heading," and continued down the hill at a rapid pace. We had played the game and he gave us the benefit of the doubt without punching our ticket. He could have sent us home with a failing grade. Whew!

We reached the "partisan's camp" for our last night in the woods. There was a huge fire in a clearing and food, glorious food. Potatoes and water as I remember. My partner and I bedded down looking forward to a walk of a few hundred yards to catch a bus for the ride back to Fairchild AFB.

The next morning, I got up finding my partner complaining of a bad stomach and weakness. I went to the fire, fixed hot tea for him and took it along with some bread back to our tent. That seemed to do the trick and he was on his feet packing for the ride home. We started out at the same time, and I was flagging; the last few days took the steam out of me. I found it difficult to stay up with him.

We came to a very small stream; I had problems negotiating it, but my partner bounded over it and toward the departure area. I saw him board the next to the last bus without so much as a fare thee well or thank you. I could see empty seats on that vehicle. I climbed onto the last and final bus. I don't remember my partner's name. I haven't thought of him since, until I wrote this section of my book. He was not the first Summer soldier and Sunshine patriot I knew, nor would he be the last. No loss there.

We bought a used Buick Roadmaster two-door coupe in Massachusetts and drove it to Washington. We traded the Buick for a '63 Oldsmobile Cutlass two-door coupe—white over metallic brown; loved that car. We sold the Renault Dauphine before we left Cape Cod. We purchased a Simca (Figure 51 above) two-door for me to drive to the base when I was going on alert duty. Built in France, I was lucky there was a nearby mechanic who specialized in this car.

BACK TO DAILY LIFE

[56]-Model Railroad Trestle

Our son, Mark, was born at Fairchild AFB in November, 1963.

I became a model railroader operating in HO gage. I scratch built all sorts of rolling stock with basswood and a gas-electric locomotive built with brass. I also built structures, buildings and a trestle.

For any modelers out there, my trestle was loosely based on an actual structure on the Spokane and Inland Empire tracks near Spokane (although in real life, I doubt an actual railroad engineer would build this structure on a curve and a four percent grade).

The trestle was a bear to build; I used 1/8" square bass wood "timbers" and you can see how the vertical risers or bents varied. Each time a timber crossed another, I notched (X-acto knife) both pieces so they formed a lap joint. After notching, I glued each joint. To facilitate the mass production, I first built a jig where I could form individual bents of any height. I formed the jig using a piece of plywood, larger than the tallest bent, and outlined the horizontal and vertical pieces with brads to hold the basswood and making notching them easier.

The "ground" bracing (horizontal slabs) under each half of the trestle are sheets of Styrofoam to simulate the grade on the sides on which the vertical bents rest. In HO, the size is about 33 inches long and 12 inches high; the real bridge was around 250 feet long and 98feet high.

The train crossing my trestle consists of an 0-6-0 Switcher and tender in brass by Pacific Fast Mail. The caboose is scratch build and all units carry the name I developed for my own railroad: the M, K & P—Missouri, Kansas & Pacific (the initials of my wife and children).

Several of us from a local club contributed and donated our efforts to a diorama for the Spokane Library—guess this trestle is gathering dust somewhere. Excuse this digression, but I did love model railroading.

MITO—PRONOUNCED MY-TOW

We were on alert duty, just in case…just in case WW III started and we had to launch our tankers to refuel the B-52s on their way to distant targets. To obviate the loss of aircraft if our bases were struck, we practiced getting our birds off the ground as quickly as possible—we practiced MITOs—Minimum Interval Take Offs.

Strategic Air Command required every combat ready crew to practice this feat of derring-do once a year. In real life, there would be a race to the runway to be the first one off, thereby avoiding the wake turbulence of the bird or birds ahead of you.

In practice, we were assigned a slot in the takeoff stream. This day we were third or fourth. My AC, Trous briefed me on his plans which included sliding a bit upwind behind the guy in front of us. Doing that would move us out of the turbulence generated by wing-tip vortices. Looking at the tail of the airplane; these are formed by the air coming off each wing tip and the vortex trailing from the left-hand wing is circulating clockwise, the one trailing from the right-hand wing is circulating counter-clockwise.

Bad enough for large planes, but they've been known to turn light aircraft upside down. Good thing to avoid. Normal takeoffs were straight ahead until the gear and flaps were retracted. At that point a turn to departure heading was completed and climb power set. When we planned to stay in a closed traffic pattern for multiple landings, gear and flaps up was followed with reduced power to maintain pattern speed.

MITO's were planned for a straight out departure at climb power for quite a distance allowing succeeding planes avoid wake turbulence as needed. Trous rotated our bird, got airborne, called for gear up and we could feel turbulence from the plane ahead. Trous eased our plane into a slight bank to slide clear and called for flaps up.

About that time, the ding-dong flying the bird in front of us honked the power back and started a turn, in our direction, as if he were going to stay in the traffic pattern. We were rapidly crawling up the guys tailpipes. Trous went to full power and began a turn the other way sending us through the full area of turbulence. I remembering him wrestling the control wheel as we were buffeted with by a clockwise vortex and immediately by a counter-clockwise one—and questioning the ancestry of the other pilot's parents.

We survived and I learned another lesson about flying—beware of idiots and plan accordingly. I think Trous had a word or two for that other pilot after we landed. Practice as much as possible, because when the time comes there isn't time to think and act—if the instinct and reactions don't come spontaneously, there's gonna be hell to pay.

SHORT TAILS

The KC-135 came off the assembly line with a fairly short vertical stabilizer on the tail which houses the rudder. Rudder is needed to counteract yaw in a turn and especially with an engine out. Most runways of our era were 150 feet wide.

The worst time to lose an engine is Critical Engine Failure Speed. On takeoff, that's the point where you should be able to abort or complete the takeoff. Of course, the worst scenario was the failure of an outboard engine and the yaw due to asymmetric power.

Our bible, the Dash – 1 for the plane, the last word for knowing the plane and staying alive stated something like this: With loss of an outboard engine at Critical Engine Failure Speed, and using optimum pilot techniques, you can expect to experience a 75-foot deviation from centerline.

Two problems: I ain't sure I'm going to be an optimum pilot that day and if I am, 75 feet puts one main gear in the dirt. The only thing worse than that was a further Dash – 1 statement to the effect: Higher Headquarters assumes this risk. Hey, Ass Hole, you ain't sittin' in the right seat.

A SECOND TALE OF A SHORT TAIL

Each year, every crew must fly and pass a check ride. An Instructor Pilot grades the pilot and copilot; an Instructor Navigator watches the Nav and the boom operator is graded by an Instructor Boom Operator. They can be real nitpickers, but most were fair in their evaluations.

My first check ride as an Aircraft Commander approached and I felt my crew was ready. I mentioned the "short tails" before, and they were tough to control with the loss of an outboard engine. Rudder trim helped, but we ran out of trim before the yaw was offset which required the pilot to hold some manual pedal input to keep the nose straight.

By this time, most of our KC-135s had been modified with the taller tail and boosted rudder. Engine out control was easy with those and we only had a couple of birds with the infamous short tail.

As we were flight planning for the check ride, I reviewed the schedule to see which airplane I would be flying. Crap. Our Wing Scheduler assigned me one of the short tails for this flight. Complaints did no good, so I flew the bird, passed the flight check and I'll always remember the guy who scheduled us to fly a short tail; thanks, Major B - - - - - n—you'll always be in my memory.

WHO RUNS THIS SHOW?

For the most part, navigators run the show until the receiver pilots have the tanker in sight. Tankers orbit at a specified point and leave the orbit on the refueling heading when the tanker and receiver navigators agree it's time. The refueling airspeed is 255 knots and the receiver approaches from 500 feet below refueling altitude—usually 25,000 feet. The receiver follows a schedule of descending airspeeds, higher than refueling speed.

Depending on receiver pilot's eyesight and weather, the receiver should have the tanker in visual contact at two miles or so. the receiver navigator, using his radar, calls our distance to the tanker to his pilot for the entire approach. At about a half-mile, the receiver will bring the rate of closure to zero and calls the Boomer. If the tanker boom operator has the receiver visual, he is in charge of the next steps. The receiver pilot has to control his plane, but the Boomer makes the calls. His next command is "cleared to pre-contact position" which is slightly below the tanker perhaps fifty-feet behind the boom tip. These calls are always preceded with the other plane's call sign then the command. Next the Boomer calls, "Cleared to contact position."

The receiver pilot will move his plane into the center of the boom envelope and call "stabilized." The Boomer will extend the boom tip and engage the refueling receptacle at which time the Boomer and receiver pilot will call "contact." Sounds easy, doesn't it?

At Castle AFB we met our air refueling instructor. Word had it, he was, in his childhood, an actor in the Our Gang movies. No matter his background, he was damn good on the boom. He was assigned another student besides me. I believe it was a fellow who became a friend and lived near us in Alaska.

My turn in the left seat came; this was the beginning of air refueling training—learning how to safely move in behind the KC-135 tanker—park it 38 feet down and below the belly of the tanker and take on fuel. Our plane was a modified tanker with a refueling receptacle mounted centerline, on top of the fuselage just behind the pilots' heads.

STOP IT SOMEWHERE

My instructor moved within a mile of the tanker and parked it a little off centerline and said, "It's all yours." So began my first lesson in what not to do. I started a small bank to move directly behind the tanker. I overshot to the other side, and tried to bank the opposite direction—but I overshot again. Soon, I was drawing horizontal S's in the sky; swinging back and forth behind the tanker. My instructor spoke again, "You don't have to stop it behind the tanker, just stop it somewhere."

Lesson learned: fly the tanker aircraft like an attitude indicator. Keep my wings level with his and the oscillations will stop. The closer I got to the tanker, the easier it was to see the tanker's wings—if I keep my wings level with his, I will stay with him whether in straight and level flight or in a turn or in a descent. Both of those latter movements came in handy on later flights. For now, we were restricted to refueling while straight and level.

WE'RE HERE ... HOW DO WE STAY HERE?

While on the boom, our instructor demonstrated the boom limits or the air refueling envelope to us—all six directions—up/down, in/out, left/right. When in contact, the tanker and the boom offer a number of clues to keep the receiver in the proper position. The boom itself has various colored tape telling about boom extension.

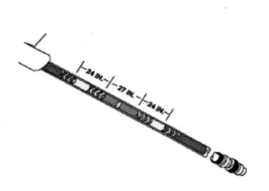

[57] Refueling boom markings

The length of the boom is 38 feet retracted and 48 feet when fully extended. We didn't have that much to play with.

In Image 57 the green extension area is 25 inches, the two yellow (caution) areas are 24 inches each. The total length we had to play with is just over 6 feet total. These markings are within the visual range of the receiver pilot.

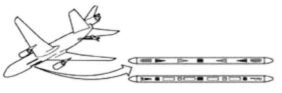

[57a] Pilot Director Lights location

A major help for the receiver pilot are the Pilot Director Lights (PDL) Figure 57b. They are located on the tanker's belly and provide UP (U) – DOWN (D) and FWD (F) – AFT (A) visual prompts.

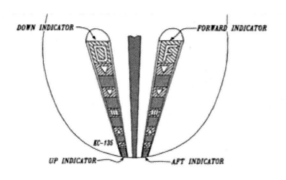

[57b] Pilot Director Lights - Cues

Figure 57c shows the PDL with vertical cues on the left and forward/aft cues on the right. The dark stripe between the two PDLs demarks the fuselage centerline and aids in maintaining position.

PUT IT ALL TOGETHER

[57c] KC-135 belly detail

The details in Figure 57d—based on the position of the tanker's centerline yellow stripe, I'm right of the tanker and need to slide left a small nudge. The double red arrows at the bottom of the image show the position of the Pilot Director Lights panels. Details of these panels are shown in Figure 57c—D/U on the left and F/A on the right (indicating: Down/Up – Forward/Aft).

The red arrow near the left center of the picture points to the bulge referred to as the boom pod. The Boom Operator is lying prone on his "ironing board" facing the tail of the bird. Just left of that arrow you can see the boon being lowered; the two black items forming a "V" are the ruddevators. The Boomer uses these to control or "fly" the boom using his "joystick" on his refueling panel.

The boom itself is partially extended. The red and yellow tape on the boom is barely visible. The receiver in this picture is nearly at the Contact position.

Later at Eielson AFB, these two camera shots are from a plane I was on—don't remember if I was in the seat for refueling or took the pics from the jump seat.

[57d] RC-135D in Pre-Contact position

Figure [57d] is the view from the pilot's seat of an RC-135D in the Pre-Contact position with a KC-135.

The boom tip will disappear from the receiver pilot's view as the Boomer extends the boom to reach the receptacle above and behind the pilots' heads.

In Figure 57e we're stabilized in the Contact position. The Boomer shoves the boom tip into the receiver's receptacle and both planes get a visual light signal the contact is solid. Both planes will verbally call and confirm the contact.

The boom operator will direct the tanker pilots to begin the refueling process by turning on the pumps from the AR panel on the console between the pilots.

[57e] RC-135D in Contact position

In training, the onload would be 5,000 pounds (750 gallons); on operational missions the fuel transfer could range up to 20,000 or 30,000 pounds. At a transfer rate of 6,500 pounds per minute (975 gallons per minute) our time on the boom wasn't unreasonable. Although, at times there were other factors we had to contend with.

The other student pilot and I did well enough on our check ride. Minimum requirements were: complete a rendezvous with the tanker, move into contact position, hold the contact for five minutes and take on a few thousand pounds of fuel. For practice, we often disconnected, moved back to the observation position and came in again to the Contact spot . We actually had contact with the boom but no fuel was transferred—we might log: one wet and two dry hookups. We both passed the check ride and Pat and I and our children were off to Alaska.

WHEN ALL ELSE FAILS

One last procedure to cover. Most of the time all goes well, however there are occasions when our best efforts don't keep us out of trouble—overrun, excessive closure rates—constituted an emergency. In that case, either the tanker boom operator or the receiver pilots can call for an immediate separation—the call is: Breakaway, Breakaway, Breakaway.

The best way to avoid this type of problem is to follow procedures, stay alert and anticipate problems. Here's an avoidance example: at Eielson AFB, my crew was on a training mission and a wing colonel decided he'd ride along. Then, he commandeered the refueling and climbed into the left seat. Since I was not an Instructor Pilot, and my copilot was sharp, I decided not to take the right seat. I did occupy the jump seat, just behind the pilots.

Rendezvous complete, the colonel began his approach to the pre-contact position. I was concerned because he was carrying a few more knots of airspeed than normal—I leaned forward in my seat. The colonel failed to stabilize before closing the last few feet. I was reaching for the throttles when my copilot clamped his left hand on top of the colonel's hand and jerked the throttles to idle providing separation from the tanker.

It takes guts for a captain to override a colonel and in essence tell him he's screwed up. Told you he was sharp. We did complete the refueling with no other comments exchanged. On return to the base, the colonel and my copilot each made a full-stop landing and takeoff again. They both made good landings and I figured I might as well get one in as well. I exchanged seats with the colonel, flew the pattern and turned final.

I didn't like to waste runway by holding the plane off the ground to get a smooth touchdown. There's an old saying the three most useless things in the world are: altitude above you, airspeed you don't have, and runway behind you. Don't never want to run out of airspeed, altitude and ideas at the same time. Anyway, since we had 12,000 feet of dry concrete ahead of us, I did waste a bit of runway to kiss it onto the concrete.

A day later, our squadron commander asked the colonel how the ride with our crew went. No one had mentioned the refueling glitch. The colonel replied, "Ah, it went okay. Both those damn pilots got a better landing than I did." Like they say, any landing you can walk away from is a good landing. More on that one later when we look at the Tale of the Bouncing Ball (Appendix B).

21) April 1967- August 1967, 16C Farewell Street, Fairbanks, Alaska

We drove to Seattle and turned our car over to the folks who would get it to Alaska—by a slow boat to Anchorage. I think this is where we had the Olds Alaskanized. Cars don't operate well at 20 to 30 degrees below zero without a few extras. These fine folks added a battery warmer (a warmer plate to sit the battery on), a circulating water heater (kept the liquid in the engine cooling system water at a reasonable temperature) and an interior electric heater. The wiring for these three elements were combined into a single lead which protruded from the car's grill as a plug. An extension cable was included which could be run from the grill plug to a hitching post. Like tying up your old cayuse outside the local saloon.

I coiled this extension cord and draped it over the radio antenna; yes, back in those days, the antenna protruded from the body. Many stores, businesses and buildings offered a source of power to keep vehicles from freezing and cracking the engine block. Each unit in our apartment complex had a plug in; and on-base housing units had their own outlet.

Hitching posts were everywhere, but never offered enough plugs to accommodate everyone who needed one. If my stop was a very short one, I would shut off the engine, dash into the store and hope I got back before the slush in the cooling system could turn solid. If the stop was of longer duration, the only answer was to cruise the parking lot waiting for someone to unhitch his pony and drive away.

SHAKE, RATTLE AND ROLL

Our apartment was furnished with scrounged furniture from Base Housing. We each had our very own metal folding army cot—with horizontal springs. We had a enough chest-of-drawers to go around. Pat and I had a good size mirror, about two by three feet, which sat on our chest-of-drawers and leaned against the wall. Empty moving boxes afforded extra storage and side tables.

Our personal furniture was stored in a Fairbanks warehouse since we didn't have room for it. Turned out to be a fortuitous event.

Good Friday, 1964 was the big one. A severe earthquake struck near Anchorage. Our fun day was a Friday, July 21, 1967. The shaker hit in the early morning hours around sun up. This dude measured in at 7.0 on the old Richter Scale; centered a few miles south of Fairbanks and lasted what seemed like hours. We were lucky, because the actual duration was only about thirty seconds.

The shaking woke me, and I watched the dresser mirror do a two-and-a-half double gainer off the dresser and land flat on the floor. Scratch one government issue mirror.

I yelled for the kids, and ushered them out the front door. I was in my skivvies as I opened the door; I waved to our neighbors and said good morning. There was little damage from the quake except for frayed nerves. Then the aftershocks began.

They lasted for days and again we were fortunate. None were of serious intensity. An aftershock is tricky. I was driving up to the Base Exchange as I saw a crowd piling out of the building running to—who knows where. I parked my car and asked a runner what the problem was. Shouting "after shake" over his shoulder, he kept running.

Why didn't I feel the shock? The only reason I can think of is the shock absorbers on the car did what they're supposed to do – absorb shocks.

Only been in Alaska a few months and already endured an earthquake. What else could happen to us? Hang onto your hats, there's way more to come from Mother Nature.

THE ALASKA CENTENNIAL

[58]-Main entrance to A-67

The 1967 Alaska Centennial celebrated the purchase of Alaska from Russia, referred to as A-67, was a fun time.

Pat and I agree that Alaska was one of the top two best tours during our time in the Air Force, so you get to enjoy more pics of this era.

Figure 58 shows a local bus in front of the main entrance.

[58a] Overall view of grounds

An aerial view of the grounds.

Figure 58a (copy of a postcard we bought), displays the overall layout of the park. They dug a channel from the river to the A-67 site so they could float the stern-wheeler Nennana (center left) in as a part of the exhibit.

A small steam locomotive and a couple of tram cars provided transportation around the circumference of the entire park.

At the left in Figure 59, the back side of the entrance sign is visible. the train is chugging from right to left toward the trestle over the pedestrian pathway into A-67.

[59] - A 67 Train

In the foreground the are a half dozen ore buckets. Dozens hung in tandem from the front arm of a gold dredge. As the dredge moved forward, the scoop buckets picked up the earth and dumped it into the center processing section of the dredge. These behemoths were about the size of a football field. When the gold ore played out, the dredges would be abandoned and left to rust. It wasn't economically feasible to recover them. The mining companies simply build another one at a new location.

[59a] Abandoned gold dredge

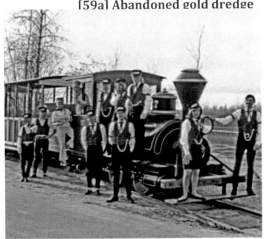

[60] - A67 Train Crew

Between runs around the park, the crew stopped the train and posed for pictures by all us "tourists" with cameras.

After a long day at A-67, this was a fun time and energy saver helping us get back to the main entrance.

There was a real fun aspect to the celebration for our girls. Each Friday during the festival, the schools allowed the students to wear "frontier" clothing.

[61]-Pat and children at A67

In Figure 61, (l to r, Kathy, Karen, Mark & Pat) pose in front of an Alaska Railroad car. the girls are wearing clothing Pat made for the occasion.

[62] - Malamute Saloon

Another fun site, near Fairbanks, was the Malamute Saloon. Probably not the same structure featured in the poem "The Shooting of Dan McGrew" by the Bard of the Yukon, Robert Service.

Notice in Figure 62, the air inside the saloon was filled with smoke. At least it was from cigarettes and not gun smoke.

On stage is the owner of the saloon reciting the Dan McGrew poem. He presented it a couple of times an evening and was serious about his performances. I witnessed the following altercation.

The owner was speaking from the stage when a lady in the audience began talking in a loud voice. The owner stopped, looked at the woman and said he was doing a serious presentation and politely asked the lady to stop talking. He resumed; she persisted; he again stopped and asked for quiet. She said she hadn't finished her beer; he said, give her the money back and throw her ass out. She wasn't injured, but she was hustled out the door.

Guess the rule is, if you own the joint, you can make the rules. I think he was right.

Alaska was called Seward's Folly when the U.S. paid $7.2 million for that chunk of land – more than 660,000 square miles. Forget about the gold, forget about other valuable minerals, forget about the huge fishing grounds and fisheries—leave all that out and the sheer beauty of the state is worth the price, which if my math is correct cost us about $11 a square mile.

It's no wonder Pat and I look back at our two and a half years up there as among our best tours in the twenty years that was an all-expense paid trip to the great white land of the Northern Lights.

THE NEXT ROUND WITH MOTHER NATURE

[63] - Pat & Fairbanks apartment

A month after the opening of Alaska 67, we were still living in Fairbanks waiting for base housing. In Figure 63, Pat is standing in front of our apartment. The steps behind her lead down to our apartment door. In the background and beyond the trees it was a short walk, downhill to the Chena River.

Pat is 5' 3" so you can gauge in later pictures how high the water rose in the end.

In mid-August, I was at Eielson AFB when Pat called and said I should come home because the water was getting close to our apartment.

I hadn't heard of any problems, but I know Pat is not subject to panic nor hyperbole. I headed for home, got within a few blocks of our place, and then I saw the water. I was able to park on high ground about a block from home. Water was creeping up the slope from the river toward our neighborhood.

Our apartment was called a "daylight" basement unit in a two-story building. The bottom rooms had windows at ground level and the rest of the room was below ground; about six to eight steps down to our front door. The property landlord, who scared the hell out of the kids, mowed the lawn, poo-pooing the idea of a flood coming. He said the river came up before but never reached his property and the apartments.

Later I learned that an old timer way up the Chena River used his radio to call down to Fairbanks with a warning. It was downhill from where he lived to our town, the land was fairly flat and the river meandered its way to us. He told the city that the water was coming and they better get out; no one paid any attention and the water caught them with their shorts around their ankles.

Good friends who lived above us had helped Pat get some of our belongings up to their place. The kids built a dirt dam about four inches high around the stair well to help keep us dry. Little did they know it would have taken a bulldozer and more earth than was available to do the job.

SIX FEET HIGH AND RISIN'

Before the water got quite that high, I made a couple of trips to the car hauling as many pieces of luggage as I could. Pretty much forgot about anything but clothing for the five of us. By this time, the water was about crouch high and I was wondering how to get my family to the car—a block away. I snapped a picture of a family across the court from us evacuating their home. A fellow in a rowboat (Figure 64) was loading them into his boat.

[64] - Escape by boat

[65] Family on the way to dry land

As soon as the rowboat departed, another fellow with an outboard motor on his boat came into the neighborhood. I hailed him, he steered toward us and moored close enough that my family didn't get their feet wet. He had a second man aboard with an oar to fend off any untoward objects.

Figure 65 shows our two saviors, (stern on left and bow with oar) in the boat and my family huddled in the middle. My apologies for the dark image, but my neighbor took the shot in haste and gave it to us later. Here we are, left to right: I am sitting in front of the boat operator, Mark is behind me in a blue jacket; Pat is behind him in yellow; Karen sandwiched just in front of Pat; and Kathy is next facing the "oarsman."

I asked our rescuer where he lived. He said he had a riverfront home and the last time he saw it; the water was up to the eaves. He figured it was a loss and the best thing he could do was help others get out. I never got his name and am sorry to this day I could not have thanked in some way beyond, "Thanks."

He dropped us at the parking lot and we climbed into the car heading for the Air Force Base. Our first obstacle was a bridge over roiling water. I seem to remember the water was sloshing over the roadway; I stopped short of the bridge; took a very deep breath and gunned the engine. We cleared the bridge

and the rest of the way was basically an easy dash to the highway and then to the base.

22) September 1967 – November 1969, Eielson Air Force Base, Alaska

September to October 1967, Visiting Officers' Quarters (VOQ) with family

[66]-Flood at high water mark

We got a two room "suite" in the VOQ (Visiting Officers Quarters) and did our best to formulate a plan. These quarters were limited to a maximum of thirty days occupancy. About a week later, I bummed a ride into town. My idea was to survey the damages and see if I could retrieve anything else from our apartment. My Ops Officer scrounged a six-man inflatable rubber raft. Came in handy.

The high-water level is visible in Figure 66 which is another picture I got from our neighbor. Remember the first image of Pat outside the building? Her head was even with those horizontal double white bars just above the water line. This was the worst of it, and when I went back the water was a good bit lower.

We did use the raft to get from high ground to the apartment to stay dry. I rescued some items from the upstairs apartment before we retreated. Another week later, the water was low enough I could walk to our unit. The stench going in was terrible; food rotting in the fridge. The water inside out place was fairly low; I didn't want to get my watch in the water. I took it off and put it in my pants pocket. Later in the day and in another building, I waded nearly waist deep in water. So much for keeping the damn watch dry.

I rescued a few pots and pans and silverware from the kitchen, and also bagged the refrigerators contents. I was gagging so badly, I considered leaving it all behind. But I couldn't do that to my neighbors.

I joined others in attempts to rescue washers and dryers from basement units. I don't know how much a washer full of water weighs, but it's a bunch. It took four of us to wrangle them up the stairs. We also brought out a dishwasher. Power had been restored to the complex, so we connected it to an outside plug. We crossed our fingers and powered it up. Lo and behold, it roared to life and ran. No soap, no hot water, but the folks could run dishes through their initial shot at mud removal.

MEANWHILE, BACK AT THE RANCH

Before I realized it, our thirty-day stint in the VOQ was up. I went to the Base Housing Officer with our dilemma. He was adamant that thirty days was the limit and we had to move. I asked him where he thought we could go. He had no answer. I gave him one: on paper—sign us out of the VOQ today, and sign us back in tomorrow—no violation of the limit and we had a roof over our heads. He didn't much care for the idea, but he got the idea I wasn't moving my family anywhere.

A short while later, my Squadron Ops Officer invited us to move into their basement. I've done my best to put the details out of my mind. We appreciated it, but did you ever live in someone's basement; a family you didn't know last week?

[67] Family and base housing

October 1967 - December 1967, 5254H Broadway, Eielson Air Force Base, Alaska 98737

December 1967 - November 1969, 5282 Coman, Eielson Air Force Base, Alaska 98737

We managed. And good luck held. A three-bedroom base housing unit opened up and we moved. We qualified for a four-bedroom and one became available in a month. Moving twice in Alaska in the last couple of months of the year ain't fun, but we were finally living in the lap of luxury.

In the picture Figure 67, (l to r) are Kathy, Mark, Karen and Pat. These were two-story buildings of six to eight units each. With a full basement, there was plenty of room.

Now the fortuitous aspect of not having our furniture in town came to be. Our household goods were in the only warehouse that didn't go under water. The local TV station had broadcast for years in black and white. Their equipment was not only dated but also ruined by water. Decision: as long as we have to replace it all, why not go with color.

Our goods included a color TV and we were among a select few in the Fairbanks area, who were watching color shows. Remember the days when we turned the set on before the broadcast day started? I can still see that picture on the screen that helped us get ALL those dials set so we could receive the best picture. That was back in the day when a show was introduced by " … and now, in living color …"

Another advantage of our latest location was a short walk across the street to the Officers' Club. In the winter, Pat put on a Saturday Dance dress, put on a parka and mukluks, tucked her high heels in a bag and we mushed over across the street. At the club, off with the boots and parka and donned the heels and we were ready for a night out. My parka was Air Force issue and was warm, but was way ugly to the mouton parkas the rest of the family wore. The same was true of the rubber soled mukluks I was issued.

Our building was one of a pair of two-story parallel buildings facing one another across the street from each other. I think ours was a six-plex while the other was an eight-plex (don't put big money on the size of that one). Our "court" also boasted a high population of children; there thirty-six as I recall. With three, we were pikers in the kid count.

OPERATIONAL MISSIONS

Flying tankers was nearly all training; the nearest to real flying were our trips to Eielson AFB and SEA (Young Tiger). Now I was in a squadron where the reverse was true. Training flights were few and far between; the vast majority of our duty was to fly classified operational missions.

The RC-135D was the basic bird. We flew missions along the Russian coast from the Bearing Sea on the east end of the USSR to the Barents Sea, half way around the earth on the west end of Russia. I flew the first few missions without the clearance to go into the back end of the airplane.

My Navigator 1 had been around for a while and I knew I could trust him. Our understanding was; if anything untoward happened in the back-end, I would take his word on whether we should abort or not. Fortunately, we didn't have to test the theory before I received all my clearances. I found out later someone was bending the rules and I should not have been flying without the clearances. Oh, well, if you can't afford a No answer…

I had an experienced copilot as well. I was pretty good on the boom for air refueling, but if I got in trouble, I knew he was a good backup. It wasn't long before I became a damn good refueler. After a few months, my experienced copilot rotated back to the real world—the lower forty-eight—and I was assigned a young Air Force Academy graduate. He was a good pilot and with tutoring he was good refueler as well.

Both my Nav 1 and Nav 2 left and I got new blood. I've always felt I was lucky, fortunate or whatever to be blessed with great men on my crews. I'm sorry ladies, but this was of an era before the contribution of women was recognized. I just mentioned two Navigators; yeah, guess they figured it took two of them to keep me from getting lost. The Nav 1 maintained our position with the INS (Inertial Navigation System) and radar while the Nav 2 used more basic tools like celestial navigation.

Since we were flying in close proximity to Russia coast the INS was our most important tool, and if it got hinky or out of whack, it was a mandatory abort situation.

LESSON LEARNED FROM AN OLD TIMER

[68] A Heading Indicator

Situational Awareness—where am I? and what is around me? —is a life saver. I'm off the coast of Russia, what's the fastest way out of here? The N1 compass system in the RC-135s was gyro driven and the display was five to six inches in diameter. I think it was graded in two-degree increments; making a rudder-turn of one to two degrees was easy.

Two considerations; what is my heading and where is the coast? The analog dial could be rotated and normally we set the current heading at the top, or 12 o'clock position. Now, I'm more interested in where the coast is. In this example, our desired heading is 000, north and the coast is off our left wing. Going the opposite direction, everything is flipped.

With each course change, I keep the desired heading pointing to my left wing. If a stranger comes up for us from a base on the coast, I make a hard right turn (90 degrees) and put the heading pointer on the nose—at the top of the gauge. My tail is now pointing at the coast and I'm heading away. A head start is our basic means of defense, and we knew their fighters didn't like to venture far from the coast. If worst came to worst, I could put it on the deck and hope.

From experience, I knew the KC-135 could drop like a rock. I brought one down one day with a rate of descent around 15,000 + feet per minute. Throttles idle and full dirty; gear and flaps down, speed brakes up full will do it.

During those days of the Cold War, most of the Russian interest was just that, interest. Seldom did they come out to see us, and if they did, no hostile intent was demonstrated. Occasionally, they did become belligerents. One of our crews had some cannon rounds fired in front of their nose. The Aircraft Commander kept his cool, stayed on course and the bad guys decided that sort of action would not keep us away.

ONE OF A KIND

The folks at Offutt AFB, flew world-wide as we did and were awarded a Distinguished Flying Cross (DFC) for each ten missions they completed. We also were awarded a medal for ten missions—the Air Medal (AM), somewhat lower in ranking from the DFC. For this and other reasons, I've always referred to us as the Northern Illegitimate Children (Bastards) of the reconnaissance fraternity in SAC and the good ol' Fightin' Fifty-Fifth. In case you didn't catch it, that's sarcasm.

Here's another example. The standing criteria for a DFC, was capturing the first signal from a new radar. One such radar was the Russian Hen House. The recon community even build a one-of-a-kind bird with special antennas to look for that radar. On one of our RC-135D flights over the Artic Ocean the back-end crew in my plane locked onto the Hen House. Wow! What a coup. And following along with protocol, we were awarded a medal for that single mission. A DFC you ask? No, an Air Medal. Politics strikes again. Did it leave a bad taste in my mouth? Freaking A it did.

DIFFERENT GEOGRAPHY FOR THE "D"

In June 1969, the RC-135M models (Combat Apple) were going through upgrades and needed some help to cover their missions. My crew was selected to take an RC-135D (Burning Candy) to Kadena AFB on Okinawa (now a part of Japan). This would prove to be an interesting tour—somewhere around six weeks.

Our old J-57 workhorse engines did well in the Arctic with cold temperatures even in the summer. Here at a latitude the same as Hawaii, there would be a problem. My first challenge was to convince the Wing Weenies (WW) that we needed special circumstances. The "M" had been fitted with turbo-fan engines with a ton more power than the older power plants on our plane. That meant all their "M's" could refuel in level flight; I couldn't. The flight plane called for a takeoff from Kadena with a first refueling abeam the southern tip of Taiwan, around 350 nautical miles to the south.

This first refueling would take us up to full max weight, and I knew I could not take on that load level at 25,000 feet. The only way to do it was to refuel in a descent. The WW's went ballistic; to get clearances for a block altitude

(25,000 down to 20,000 feet) would cause extra work for the staff—oh, dear me.

After a bit of sniffing and snuffing and excuse after excuse, I'd had it. I told them to schedule a training mission to simulate the flight and see how long it takes for me to fall out of the air. This proposal would take a recce bird out of the operational mission rotation and tie up a tanker as well. Reluctantly, they relented and I got my block altitude.

Our mission required a night takeoff, that first refuel shortly after takeoff, a few hours to reach the Gulf of Tonkin (GoT) where we spend more hours boring holes in the sky about 50 miles off Haiphong Harbor, North Vietnam. And to keep us airborne, they had another hurtle in store.

We headed back down to the southern end of the GoT, so we could hit two more tankers. They flew in formation and our plan was to take on fuel from the lead on the southbound leg, turn 180 degrees, and then slid over to get our fuel from the second tanker on the northbound leg.

The refueling legs were long enough to complete the onload from one tanker. I wanted to get some refueling training for my copilot, so I let him chase the first tanker southbound. We hadn't onloaded fuel yet. He did a good job, but you don't learn to hang on the boom the first time out. Later on, he became a damn good refueler.

But—at this point it was time to turn and the powers that be frowned on refueling in a turn. At the Pre-takeoff briefing, I met the tanker pilots and one was a squadron mate from a previous assignment. The tanker made the "end of track" call and I said, "I know. I won't tell if you don't." His response was positive and I moved into contact position, and the boom thudded into our receptacle. The tanker pilot called the start of his turn, and I added a bit of power and thumbed in nose up trim to offset the loss of lift in the turn.

By the time we were northbound, lead tanker's offload was complete and I moved over the Number Two and got our second briefed load of fuel. After that we went back up to the Haiphong area for more hours. When we hit bingo fuel, we headed south, exited the GoT and then north again to make it back to Kadena AFB.

With all that fuel, they kept us airborne for nineteen and a half hours of flying time. Counting pre- and post-briefings, our crew day ran almost 24 hours. We flew that schedule every third day for a month. With overlaps, we were in crew rest by the time we woke from the previous mission, which meant we couldn't even have a beer let alone have any free time. I talked to the Wing scheduler, they relented and we got a couple of days off. Time to go shopping so we could drag junk home—and enjoy a beer or two.

The night refueling off Taiwan got real dicey on one flight. Bad weather moved in and we were dodging thunderstorms. I latched onto the tanker and he began his descent when we hit an area of heavy weather. We were bouncing around, but I was able to hang on long enough to get our fuel. After we cleared the tanker, I looked over at my copilot; he said, "If I'd been flying the bird, I would have called a breakaway."

He had a point, however this was an operational mission, perhaps vital to the air war in SEA. I probably worked harder on that refueling than ever before; I imagine my flight suit was soaked with sweat. If the weather had gotten worse, I would have aborted the refueling. If I felt I was unsafe or reaching beyond my capabilities, I would have aborted. It's a fine line between getting the job done and pushing you or your bird past their limits.

One of my fondest memories of this Kadena tour was being adopted as an honorary member of F Troop. That "unit" was part of the Kadena maintenance cadre and their name came from an old TV western sitcom from the mid-1960s starring Forrest Tucker and Ken Berry. A small framed "certificate" still has a place on my wall.

CAGE NUMBER THREE

Back to the Gulf. As were exiting the mission area, we noticed fluctuating oil pressure on number three engine. We swapped gauges with one of the other engines, and verified the gauges were good, the engine was bad. We shut the engine down; not a major problem since three engines would keep us airborne and there was enough fuel to get home.

The traffic controller issued a clearance for the normal route home. Bit of a sticky wicket; the "normal" altitude going home was around 33,000 feet. With only three engines turning, we wouldn't be able to get that high. Without mentioning the engine shut-down, I told him "unable FL 330, got anything around 2-5-0?" Good controller; didn't ask questions; assigned us the lower altitude. When we were within UHF radio range, I notified our Command Post, as required, of our engine-out. Easy three-engine landing and no repercussions.

AUTOMATIC APPROACHS

The 135 had a decent autopilot; nothing like the ones today that can land the plane in zero-zero conditions, but it eased the weariness of long flights. It also had the capability to linking to an Instrument Landing System (ILS) and flying an approach in instrument conditions. We could fly it down the glide slope to around 500 feet and then manually make the landing.

At the end of one or our marathon flights to the GoT and back, I asked my copilot if he'd ever used the autopilot for an approach. He said no and I coached him through the procedures. It was a beautiful clear morning so the approach should be a good learning situation. We tuned the ILS frequency, received the proper aural identification and there were no red flags showing on the cockpit instrument.

We were level inbound when the ILS locked onto the localizer which would guide us laterally to the runway. We set up for landing and glide slope needle moved down from the top of the instrument case; when it reached the middle, the altitude hold clicked off and we started down the glide slope. During all these procedures, the pilot only has to control the airspeed with the throttles. All was going great.

We were descending and about three miles from touchdown when our navigator called out that our position on his radar was wonky. I rechecked visually and sure enough we were a half to three-quarters of a mile left of the center line. The tower called us and pointed out the same information. We called runway in sight, altered our heading to line up with the runway and the landing was completed without incident.

My attempted demonstration of autopilot approaches ended up satisfying two objectives. My copilot learned new procedures he could tuck away in his flying kit of instinctive reactions. Second, we again learned to be paranoid and hope for the best and plan for the worst. It's always good when all crewmembers are alert for problems; in this case the navigator. Flying visually that morning, my copilot and I would have seen the error and corrected. In instruments conditions, the pilots or the tower might not have recognized the problem until it was too late. Such are the cause of accidents and dead crews.

On the ground in maintenance debriefing, we described the problem. All checks of the ILS: frequency, audio signal, localizer and glide slope needles centered and no red flags showing gave the correct picture. Somehow, the instrument screwed up and ground folks had no idea why.

I checked back later to see if there was an answer to our problems. There wasn't. Lesson learned—don't trust a damn thing without double-checking.

OVER THE TOP TO THE UK

Most of the operational flights from Eielson AFB were out and backs—takeoff, refuel, spend eight to ten hours loitering near the Russian coast and go home to Alaska. A few took us so far west, the Kara Sea or the Barents Sea, that England was closer than our home base.

When the mission was complete, we would follow the coast of Norway and over the North Sea into England. In our day, the recovery base was RAF Upper Heyford, near Oxford.

If we landed on time, we could finish debriefing, check into the VOQ, change and call for a taxi. A short ride to the railway station at Bicester North and we could catch the last train to London. We usually made it. We stayed at the Columbia Club on Bayswater Road across the street from Hyde Park.

We used the tube, the London Underground, most of the time. It got us near the Tower of London, Buckingham Palace, Parliament, Hyde Par, and Big Ben. On Sundays it was a short walk to a fence on the north side of Hyde Park where local artists displayed oil paintings. I think we all brought home at least one painting.

[70] Hyde Park

Figure 71 shows the sentry guarding the path to the bridge and the entrance to the Tower of London.

[71] Entry to Tower of London

The changing of the guard at Buckingham Palace, London.

[72] Buckingham Palace – Changing of the Guard

Figure 73 indicates the area where unlucky nobility lost their heads. The post and chains demark the area of the scaffold.

The acatual chopping block has been removed.

The list of the headless victims are shown below.

[73] Site of Execution Scaffold

The sign behind the area reads:
ON THIS SITE STOOD A SCAFFOLD ON WHICH WERE EXECUTED
    Queen Anne Boleyn   1536
    Margaret, Countess of Salisbury   1541
    Queen Catherine Howard   1542
    Jane Viscountess Rochford   1542
    Lady Jane Grey   1554
    Robert Devereux Earl of Essex   1601
also near this spot was beheaded Lord Hastings 1483

On outings around the air base, a short cab ride would get us to a tavern or inn. After calling ahead, most inns serving dinner were small and took reservations, we hired a taxi for our night out. This was in the days of crowns, farthings, halfpennies (hey-pence) and the weird exchange between the Pound Sterling and the U. S. Dollar. To ease our pain, each would chip in a five-pound note, give it to one person in the group and it was their headache to pay for everything,

Mostly, when the cabbie announced the fare or a waiter presented a bill, the person with the money would hold a batch of currency in cupped hands with an expression that said, "don't hurt me too bad." I don't think any local ever gouged us.

One day, bored stiff, we called the local taxi service. The lady asked where we wanted to go. Our answer: charge us by the hour and take us on a sightseeing tour of the local area. She thought us a weird bunch of Yanks, but did a great job.

[74] Blenheim Palace

We passed Blenheim Palace where Winston Churchill was born and buried. It has a long history as the home of the Duke of Marlboro.

[75] Sir Winston's grave

Not too far away is Churchill's burial place. It's the graveyard associated with Blandford Church.

I believe Sir Winston's grave is the second stone up in the lower left of Figure 75.

[76] Rollright Stones

Our lady driver also took us by an area of stones. Nowhere near the size of the ones at Stonehenge, but with an interesting history. I always found it interesting that Americans think in terms of a country as a couple of hundreds of years old while the Brits think in terms of centuries.

The road sign at the entrance, Figure 76 the Admission price, Free, and announces the stones are referred to as the King's Men. This area dates from the Bronze Age around 1500 BC.

Figure 77 – visitors walking the circumference of the stones.

Folk stories of the area say a person cannot walk the full circle counting the rocks and come up with the same number twice in a row.

Figure 78 is a large stone, a monolith a short distance from the circle.

[77] Visitors at the circle of stones.

[78] King Stone

The sign in front of it reads:
THE KING STONE
A MONOLITH PERHAPS
ADJOINING STONE CIRCLE.

We were all disappointed when the tour ended. The country side was beautiful, tranquil, and interesting to learn about. But back to work.

Robert Service, the Bard of the Yukon, paid tribute to northern lights with "The Ballad of the Northern Lights" and this line; "The Northern Lights have seen queer sights, but the queerest they ever did see…" from "The Cremation of Sam McGee." They're beautiful dancing over the skies, and they gave me fits on a night refueling.

We flew west from central Alaska, and out over the Artic Sea towing a tanker along. We led the flight since we had the better navigation equipment. Radio silence was in effect so we briefed the KC-135 crew that we would key the UHF radio twice when it was time to descend to refueling altitude. At the proper altitude, I would level off and the tanker was to assume the lead and I would slide over behind him for the air-to-air refueling.

One other night, I was carrying fifteen to twenty knots more airspeed and not closing on the other plane. He forgot to slow to the proper speed; had to break radio silence to get him to slow down.

Back to the Northern Lights. This night, all went well and I was closing on the tanker when the lights began their dance. They formed ahead of us and looked like a horizon when the sun is coming up. The bad part was, this horizon was slanted about twenty degrees off level. My eyes played tricks with me and made me "feel" like we were in a turn.

I fought it off and we completed our onload with no real problem. As soon as I disconnected from the boom and lost some altitude to clear the tanker, my lead navigator went on the radio with a short blurb, like "twenty-five short." That told the tanker navigator we were on track and twenty-five nautical miles short of the planned end of refueling track point. He had a positive fix to start his trek home.

HOW TO SHAKE UP THE BRASS

On one of the missions off the northwestern Russian coast, we experienced a failure of some mission essential equipment. We weren't in any real danger, but a mission abort was required. Also required was an encoded message to be sent back to Eielson AFB to appraise them of the facts.

We used our codebook, wrote down the characters we would transmit and put the message out over the HF (high frequency) radio. Somewhere in the preparation, I transposed a couple of numbers about the abort longitude.

Our plan for the abort was to head north toward the north pole and then turn to a line of longitude which would take us to Alaska. The numbers I transmitted indicated we would overfly Russian territory on our way home. Our actual route was well clear of their territory, but our guys on the ground would not know that until we landed and debriefed.

Keep in mind that all lines of longitude (the ones running north/south on a map) all around the world converge at the north pole. When we neared the pole, I went back to my Nav 1's station to watch the inertial nav system at work. A small digital dial kept track of the longitude along our track. We came within ten or twelve miles of the pole and the lines were so close together the digits on that dial were a virtual blur.

The rest of the flight was uneventful until landing when the commanders came looking for the reason we overflew the USSR. A quick check of the navigator's log clearly showed we didn't. When confronted with our decoded abort message, I saw why they were upset. Instead of creating an international incident, I admitted to the screw-up.

THIS IS COMMUNICATING WITH THE TROOPS?

I was at work one day, and a neighbor of mine passed me in the hall and congratulated me. I asked him "what for?" He said I just made regular Air Force. Up to this point I was a reserve officer on active duty and apparently had been added to the rolls of the Regular Air Force. Bruce said he read it in Air Force Times newsletter.

I checked with our wing personnel office and they verified the change in status as being correct. I asked why I had not been notified and got one of those eye-roll shoulder shrugs. Keeping the troops in the loop it wasn't.

THE OTHER HALF OF OUR EIELSON DUTIES

In addition to the three RC-135 "D" models stationed at Eielson AFB, the wing owned a pair of one-of-a-kind airplanes which spent most of their life at Shemya Air Station, Alaska. Shemya is a dot in the ocean at the far end of the Aleutian Island chain. A section of the islands is called the Near Island—guess because they're near Russia. A smaller group, three islands, the Semichi Islands was our home. Agattu, a larger island is southeast, to the west is Attu.

Shemya is covered with WW II buildings and they show their age. None are painted, since the harsh weather and wind would scour them bare anyway. Two large hangers housed our two special birds. Rivet Amber, was an RC-135E and carried what must have been the world's largest airborne radar. I wasn't qualified to fly that one, but I flew the RC-135S, Rivet Ball. Ball sported a row of ten windows down the right side of the fuselage. Each window was about eighteen-inches in diameter and was made from optical quality ground quartz. Our mission was designated as PHOTOINT, so you have an idea of what the windows were for. (I discussed Rivet Amber earlier as well, and my memoir "The Iron Pumpkin" covers River Ball in detail).

CHECK AND RECHECK

On arrival at Eielson, I had a "differences" checkride to learn the RC-135D. Got another check for refueling and then another for night refueling. Before flying the "S" model, another check was in store. First, understand that Shemya has been referred to as the asshole of the world as it relates to weather. I once said: If the good lord wanted to give the earth an enema, the insertion point would be Shemya.

An Instructor Pilot flew the right seat in a KC-135, borrowed from the Eielson Tanker Task Force (ETTF) and off we went. I was required to make three full-stop landings and takeoffs at Shemya in the tanker. Didn't want to risk one of the special birds. The tanker version of the 135 cost around five-million dollars. By the way; back in the 60's the "D" models came in around $40 million; the "S" was worth about $60 million while the "E" was king of the hill at $80 million. That was a pile of money back in the day, but compared to the price of airplanes today, barely pocket change.

Meanwhile, back on the runway. To become "island" qualified, not only did I have to make the landings, miserable weather was required. As I recall, we had to have at least a 50-degree crosswind wind with speeds over 20 knots. Not a fun day and it falls in the sweaty flightsuit category.

For fun and pictures, see Appendix B about the Bouncing Ball.

Tanker pilots were not allowed to land at Shemya; an Aircraft Commander from our squadron had to go along and make the landing. I flew there with an ETTF crew one day. I told their crew commander to log all but five minutes of flight time to him and his copilot; and he could log five minutes for me as Aircraft Commander time for the landing. He asked why I didn't just log Instructor Pilot (IP) time for the whole flight. When I told him I wasn't an IP, his jaw dropped. I made the landing and takeoff and slept most of the rest of the flight.

FYI – example: on a five-hour flight, four pilots could each log five hours of flying time. The regular pilots could log pilot or copilot time. A third pilot who is an Aircraft Commander could five hours AC time. And if a fourth pilot was an Instructor Pilot; he could get IP time for five hours. Rather a weird way of tracking flying time, but remember this is a military organization.

HOME AWAY FROM HOME

Front end crews spent a week at a time every six or seven weeks; not too bad since we had a dozen crews. Our back enders consisted of two teams. They spent half their Eielson tour on the Rock.

These teams were crewed by Electronic Warfare Officers—EWOs—who at times referred to themselves as Ravens or Crows. The bird on the Team 2 patch is a raven. I became an honorary member of Team 2—but more about that later.

[79] Team 2 Patch

In Figure 80, you can get an idea of the size of the hangers. It shows Rivet Ball completely inside the hanger. Add a few feet to make way for the tug—wheel barely showing at the right—and the yellow tow bar attached to the nose strut. When a launch signal came, one tug pushed the huge hanger doors open and the other tug pulled us out of the hanger.

A stairway—roughly two stories high—led from the hanger floor up to the level where the crew quarters were located.

R & R

To while away the time between alerts, we had a ping pong table on the hanger floor. It that got boring, we created paper airplanes, climbed half way up the stairs and played carrier landings with the ping pong table as the carrier. We also had a volley ball net set up on the hanger floor just in front of the right wing of the plane.

A DAY IN THE LIFE

[80] The Ball in the hanger

The hanger had no food service. We piled onto a crew bus three times a day for meals.

There were no civilian vehicles on Shemya. A couple of GI buses hauled permanent residents around the base. Guess this is a good time to relate the tale of the dogs on the island. The lead dog, Boozer, was a big black and white husky who hung around the headquarters most of the time. I'm not sure where his name came from, but you can guess as well as I.

[80a] Boozer resting

Figure 80a shows the old dog resting on the concrete steps outside the headquarters building.

Two or three other non-descript pooches lived on the island as well. When they needed to get to another part of the island, they used the only transportation available. They took the bus. They would wait at a regular stop and the driver would open the door—they hopped on and waited till the driver opened the door at "their"

[80b] Boozer's Plaque

stop and climbed off. I think their favorite destination was the one nearest the mess hall.

Old Boozer was such a beloved part of the island, when he died, they had a memorial service outside the headquarters building. I don't know where the old guy is buried; I'm pretty sure there is a gravestone and I know a plaque to him was placed outside headquarters; it reads:

"Boozer" 31 December 1968
The Greatest Morale Factor on Shemya since WW II

Hell, I'd accept that tribute in my own obituary.

The USSR had three launch sites for intercontinental ballistic missiles: Plesetsk, Kapustin Yar and Tyuratam. Although these sites were widely spaced in central Russia, all three were aimed at a single target. That's why we spent our time hanging around the Kamchatka Peninsula.

For more details about the Rivet Ball, read my memoir "The Iron Pumpkin," which is my story of me and the bird.

IS IT DANGEROUS?

From time to time, folks have asked me if I was afraid of what I did for twenty years. After all, flying can be dangerous. Anxious, anxiety—perhaps. I have described myself as being paranoid trying to figure out who or what might b e trying to kill me. A bit like, plan for the worst and hope of the best.

To alleviate those thoughts, activities like training and practice are great preventatives. In personal matters, I've used an anxiety reducer. Set up the situation in my mind; imagine the absolute best circumstances and outcome; then imagine the absolute worst circumstances and outcome. Now I've faced the worst and the best that can happen; and the actual outcome will be somewhere in between.

If I were to think, today's the day we're going to (fill in the blank). I doubt anyone would do what we do. I was as prepared as I could be to face whatever came along. Many times, all the preparation was not needed. A few times it was and the outcome was satisfactory.

A TRIP TO THE MIDDLE OF NOWHERE

Early on a Friday, we got the word from SAC Reconnaissance Center (SRC—Offutt AFB, Nebraska) that we should get Ball ready to deploy to Johnston Island—around 500 miles west-southwest of Hawaii.

Seems the French decided to resume atomic tests in the South Pacific. The target would be Moruroa atoll, 2,000 miles southeast of Johnston Island. Since it was a high priority mission, we expedited our mission planning to Johnston with a RON (remain overnight) at Hickam AFB, Hawaii. We told them we could launch by about 3 p.m. that day.

The local time in Nebraska was past duty hours there and, in the Pentagon, who's okay they needed. SRC wasn't prepared for our fast response, and after a couple of sputters, the directed us to depart on Monday.

A dedicated KC-135 and crew was assigned to accompany us. I don't remember where he came from, but we had him for the duration and they landed on Johnston Island with us. Having deployed directly from Shemya, our suitcases contained a shaving kit, a week's worth of underwear and a spare flightsuit. We hit the Base Exchange in Hawaii and on the island and bought out their stock of Hawaiian shirts and Bermuda shorts. Our attire didn't do anything to put the island commander (IC) on our side. At the time of our arrival, the U.S. Navy owned the island, but somehow the Army was in charge.

We were on a short leash for takeoffs and needed transportation to our plane. This didn't sit well with the IC either; especially since we might have to takeoff at night. The IC, thinking he had a homosexual problem on his island, had prohibited driving at night. Never did understand how a night driving ban could curtail anything. Messages from our headquarters arrived, he changed his mind and we wore our shirts and shorts and drove at night if need be.

[81] Johnston Island

The island is small, a bit less that a square mile in area. As seen in Figure 81, there isn't much dirt beyond the runway in any direction. As for length of the runway, the best I can find now is that it was 9,000 feet. The place's best attribute was the civilian run mess hall. Once a week it was steak night; on another week night they served prime rib. These meals were served on large oval steak platters and the meat lapped over the edges—fantastic fare.

FUN IN THE SUN

Island entertainment consisted of an outdoor movie theater and shark fishing. One day we came upon guys with large hooks at the end of piano wire—and a shark in the bed of an Econoline Ford pickup. The shark's nose was in the left front corner of the bed and its tail draped over the right rear corner of the tailgate. A fellow was sitting on the right edge of the bed with the shark's back toward him. He jabbed the beast with something and the shark gave a shudder—the guy in the bed did a backflip off the truck.

The outdoor movie theater was close to the runway, as was everything on the island. I remember a late evening takeoff and seeing our wing tip almost reach the theater bounds. I understood at one time this garden spot was an accompanied tour for naval personnel. Once a month a flight took spouses to Hawaii for shopping and to offset the isolation of the assignment.

[82] Crew E-10 at PJON

My crew, E-10 at Johnston Island (PJON). Figure 82 From left to right: the author; Usto (Sam) Schulz, copilot; Ellis (Stu) Williams, Nav 2 and Paul Brown (Nav 1).

In the beginning, I asked what we would be doing out there. The answer included the French may be testing atomic weapons; just fly around in big circles until you glow in the dark—then you can come home. The Operations Order for coverage of the tests was called Burning Light. This name pops up later as well.

BACK TO WORK

Normally, we wore our brain buckets (crash helmets) during air refueling. I tried that on the first mission—big mistake. In that South Pacific weather, the sweat poured down my forehead like Niagara Falls. Thereafter, we passed on that requirement. The AC of the tanker tagging along was a good stick. We talked on the ground and he knew we would need to refuel in a descent to get full tanks.

He was smooth as silk, and he flew a platform that was a dream to refuel from. After passing his gas, he headed back to Johnston Island, and we set out for a French protectorate of Moruroa atoll; 2,000 miles to the southeast. We spent a few hours orbiting the area until we hit Bingo fuel and went home without any results.

This process was repeated several times, and apparently the French decided to call off atomic tests. Didn't gather much in the way of intelligence, but it was a respite from the arduous weather in the Aleutian Islands.

MEANWHILE SHEMYA CALLS US HOME

[82a] Hanger bedroom

Before we get to the fun and games, I'll show off the sumptuous quarters on the upper level of the Ball hanger. More like Early GI Bleak, at least it wasn't Khaki. This is half a two-man room.

In Figure 82a, I'm sitting in my bedroom. This picture is one of a series I shot of myself. Camera on a tripod I was testing the lens settings on a new 35mm SLR camera. Not quite the selfies of today.

The time between flights could become boring; most of us had hobbies or habits to help while away the time. And, then the Viper struck (I think it was him). We were expecting an Inspector General (IG) team to inspect our detachment the following day.

The morning of...here are a couple of pictures of Rivet Ball in all her glory that greeted our Det Commander.

[83] Ball the Shark - distant

Ever hear of the Flying Tigers of WW II?

Look closely at the shiny thimble nose for the latest additions to Rivet Ball.

[84] The Ball's shark nose - near

The black nose was an expensive add-on and vital to the clean operation of the enclosed radar. The shark eyes and mouth sure seemed to be painted on, which would require a ton of money to replace. Our Detachment Commander nearly had a S _ _ t Hemorrhage,

Our culprits had the good sense to use paper and very small bits to tape to add the nose art. Much relieved, the Det 1 Commander, left the things in place and the IG team leader had a sense of humor and a laugh. All involved skated free and we passed the inspection.

ONE LAST FLIGHT

Late on January 12, 1969, we launched on a higher headquarters directed mission. We returned to the Shemya area around midnight and were cleared to land. At thirty-minutes past midnight on 13 January 1969 we landed on a runway—unknown to me as the pilot—covered by slush. We hydroplaned off the end of the runway and over a forty-foot cliff.

The airframe was a total loss; salvage recovered one of the four engines, the tail section and around 60,000 pounds of electronic gear.

There was no crew error assigned to the accident. Weather and operations personnel caught the blame.

THERE WERE EIGHTEEN CREW MEMBERS ABOARD ...

WE ALL WALKED AWAY!

All the details are too much for this journal, I've included them in "Rivet Ball, My Story" aka "The Iron Pumpkin."

BEFORE WE BID A FOND FAREWELL …

[85] SYA bleak and barren

Bleak and barren would be a good description of much of the island. Figure 85 shows some land near the beach.

Large concrete "plug" with a rusty chain attached to a U-bolt in the plug. I recall it as being near the headquarters building.

I don't think anyone knows what this is—Figure 86. Rumors have it that if the Plug is pulled, the whole damn island will sink, hence the "sturdy" fence for protection.

[86] The Plug

This may be the weirdest. The bridge to nowhere. Again rumor says some went berserk on Shemya and began to build a bridge to the next island.

[87] The Bridge

Another aside a daughter reminded me of. Back at Eielson.

Behind our six-unit building, there was a depression I'd guess was about 30 yards by 10 yards and a foot or so deep. Just before the winter freeze, the Eielson AFB Fire Department would pull a pumper truck up to our backyard and reel out the hoses.

The freeze would hit, and we had our very own skating rink. Pat and the girls did a creditable job on the ice. I think Mark was a little too young, but I think he gave it a shot as well. Me, I was really good at lacing up the skating boots, but standing up for any length of time was a totally different challenge.

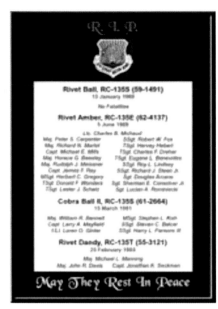

This plaque, 6SW RIP, lists four aircraft losses—from the second one down: Rivet Amber, 5 June 1969; Cobra Ball II, 15 March 1981; and Rivet Dandy 23 February 1985. There was loss of life on all three.

I'm proud of the first listing: Rivet Ball, 13 January 1969 and the words: NO FATALITIES.

[88] 6SW RIP Plaque

A FINAL — FINAL WORD

[89] A stripped carcass

Parts like the air refueling receptacle, cargo door, engines, and tail were stripped leaving some gaping holes. This is the original resting spot after the crash. The carcass was finally moved to the base dump a little west of this position.

As last tribute, I'm including the poem I wrote in 2009. "We All Walked Away."

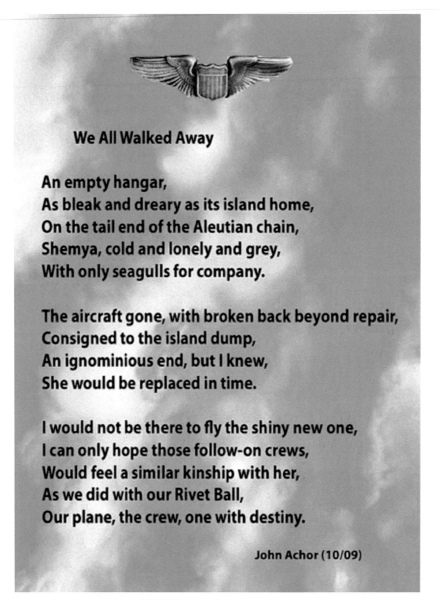

[90] Empty Hanger poem

This may be a fitting time to depart Shemya for the last time.

23) November 1969, 2 Carter Terrace, Daytona Beach Shores, Florida 32018

MY LUCK HOLDS

Following two and a half years in Alaska, I received a southern assignment—to the ROK (Republic of Korea), aka South Korea.

Before reporting overseas, again, I did get the opportunity to move my family back to the CONUS. Pat's parents were living in Daytona Beach, Florida and it seemed a good temporary location.

A LAST OUNCE OF FLESH

Our bags were packed, as were our household goods, and we were on our way—and then. My scheduling officer said they needed one more flight from me—a couple of days after Pat's flight out on Alaska Airlines. We decided she would go ahead without me; not that the situation was a first. The biggest problem; that left her to shepherd our three children from Alaska to Florida.

On my solo flight home, I was seated behind a couple of Great White Hunters returning to the states from a hunting trip to Alaska. I overheard one tell the other, "Just before I needed to leave, I went out one more time and heard something in the brush. I took a sound shot and went to the site. I found a blood trail, but it was time to catch the ride to the airport, so I quit."

"Sound" shot? Blood trail; wounded bear roaming the woods and really pissed off. I felt like strangling the idiot. I didn't because that would have delayed my return to Florida and my family.

The one good part of the PCS (Permanent Change of Station) move was a school I would attend before leaving for Korea. That gave me the authority to head back to the CONUS, get the family settled and attend the school, and have my personal belongings crated up before getting on my jet plane.

My school, Tactical Reconnaissance Officer training was at Myrtle Beach AFB, South Carolina. I completed the course but still didn't know much about TAC Recce, but the square was checked and I could be one in Korea.

NEED NEW TRANSPORTATION

We figured two winters in Alaska, took their toll on the Oldsmobile, so we decided to buy a new car on the way home. We left the Olds behind with a dealer on consignment. Now for a new one.

The only 1970 model of any type or brand for sale at the time was the Ford Maverick. A bit larger than the Mustang it was also a two-door coupe. I think we took possession of the car somewhere in Florida when I finally made it there.

ONE LAST ALASKA TALE

This next section was previously published in my memoir "The Iron Pumpkin," but I must include it here as well.

A TRIBUTE TO LURCH

Somewhere on my computer's hard drive is a story about Al Hansen. I think perhaps Al, with a mischievous smirk on his face hid it from me, so I will attempt to recreate it here. Lurch was the first of the Team Two, Ball crewmembers from that night to die, it is only fitting that I give Al this final story in my book about Rivet Ball and let him have the last laugh.

Alan Hansen was a photographer and an Electronic Warfare Officer; what better place to serve the Air Force than on an airplane that took momentous and unique photos that may have well changed the world. Al picked up the crew nickname Lurch based on his size. He carried the largest frame on the crew.

He was one of the other seventeen guys who rode the Rivet Ball off the cliff with me at Shemya Air Base on 13 January 1969. That desolate and remote island at the end of the Aleutian Island chain was the "home" for the RC-135S, 59-1491.

The plane was a total wreck lying there in the dirt at the bottom of a 40-foot drop. We all suffered through the accident investigation board and finally made it back to Eielson Air Force Base. With the loss of this one-of-a-kind bird, Al's job melted away. He rotated back to the CONUS and I followed a few months later.

While at Offutt Air Force Base and filling a ground job, I flew the T-29, a navigation training plane with Base Flight. I was scheduled to fly a round robin carrying a batch of others who needed the flying time to collect flight pay. It had been a least a year since Al and I parted ways.

I was sitting on a front seat in the crew bus as the passengers climbed on. I recognized Al and we exchanged friendly nods as he passed. We completed the mission and landed, then headed for the crew bus again. I beat the passengers to the bus and took a seat up front again.

The passengers clamored on board; as Al topped the steps, he stopped next to me, and in that quiet, wry wit only Lurch could pull off, he said, "Better landing than last time."

The End

Even though this incident happened at Offutt AFB a couple of years after Alaska, I always remember Lurch on the Rivet Ball Team Two in Alaska.

OFF TO MY SOUTHERN ASSIGNMENT – KOREA THAT IS

Since I was assigned to the 314th Air Division as a Tactical Reconnaissance Officer I was slotted to Tac Recce school at Myrtle Beach AFB, South Carolina. It was only two weeks long, but that got me back to the CONUS, so I could get Pat and our children settled into a yearlong rental. We found a great three bedroom house in Daytona Beach Shores (on the south side of Daytona Beach).

I finished the ground school that would 'transform" me into a tac recce pilot and wondered if all that "expertise" I garnered would really qualify me for my new position.

THIRTEEN MONTH UNACCOMPANIED TOUR

Why thirteen months? Well, back in the day the one-year tour was extended by the Army to include two weeks travel by ship on both ends. Now here in the days of jet aircraft, the 13 months held.

I took commercial flights from Florida to SEATAC (Seattle-Tacoma airport) where I transferred to a stretch DC-8 civilian contract airline. Boy, was I lucky. The bird was filled to the gills with dependents. More wives and children than I could count. After a refueling stop in Alaska, we set sail for Japan where most of the passengers deplaned.

We continued to Seoul Airport where I got off. All the military people were greeted by a huge Master Sergeant who was barking orders like: Turn in your green, get your Won, do this, do that and on and on.

When he ran out of breath, I asked him to translate his words into English. Most of his orders were for the lower ranks and didn't apply to me. He directed me to my transport to Osan AFB, about 30 miles south of Seoul.

24) December 1969 – December 1970, Osan Air Base, Korea

My home away from home was a room on the second floor of a two-story prefab type building. An external stairway led to my floor, and I don't remember if there were stairs inside; don't think so. I was assigned to a part of the 314th Air Division as a TAC Recce officer for a few weeks.

My digs at the top of the hill (Figure 91). What seems to be a wall behind me is the back of a double wide armoire, my only closet. Behind the armoire is an area about 12 feet by 8 feet which was the bedroom. This side was about the same size as the bedroom. Behind the camera is a wall separating the bathroom area. Overall, the entire place was about 12 by 24 feet.

[91] Osan BOQ room

Note the Christmas décor. This would be 1969 since I left before that second holiday rolled around.

The BOQ was on a hill toward the west side of the base and was about the highest point on Osan AB.

I was scheduled to go the 5th Air Force ADVON (Advanced Operational Node – 5th AF Headquarters was in Japan); a one-man unit who was due to rotate to the U.S. shortly. I met the man; Jack Donovan—perhaps the most talented officer I ever knew.

Jack was an Electronic Warfare Officer (EWO, Crow, Raven, in the Wild Weasel program, a Bear) and came out of B-52s. He volunteered for an assignment blind, before the term Wild Weasel was invented and became world renown. Those were the guys who in Vietnam went in before the strike force to suppress SAM (Surface to Air Missile) sites and protect the bombers. Their motto said it all: First In – Last Out.

Donovan joined an F-100F outfit while they were designing and building a concept plane to do the job. He was to fly in the back seat of this fighter and learn to kill SAM sites. He came up with the words that became the acronym—YGBSM—which is part of the Wild Weasel (WW) patch to this day.

Here are the words Jack used to describe the WW mission: "You want me to fly in the back of a tiny fighter aircraft with a crazy fighter pilot who thinks he's invincible, home in on a SAM site in North Vietnam and shoot it down before it shoots me. **You Gotta' Be Shitting Me!**"

Their motto First In – Last Out meant they entered the target area before the bombers/fighter bombers were due to arrive and stayed on station until the last of those bombers were out of danger.

Their job was to counter the Russian built SAMs employed by the North Vietnamese. They accomplished this either by killing a SAM site, or by scaring them from bringing their Fan Song radar up on air fearing a Shrike missile (AGM-45) from a Wild Weasel would follow their radar signal straight into their radar van.

The WW story is told in a great video available on YouTube, "First In, Last Out, The Story of the Wild Weasel." I created a bitly (shortened URL to a site) to type in the browser address line to locate it: bit.ly/weasel-jack (don't miss the period between bit and ly) It's still current as of late 2022.

It is an interesting story, and there are several clips of Jack Donovan included. Jack and his pilot, in their F-100F, were credited with the first SAM kill in North Vietnam. There was an honorary fly-over when he was buried at Arlington. I'm sorry to say that, Jack, died April 11, 2015. I miss you, my friend.

Jack's mid-tour leave came up and he headed for Orlando, Florida. Unknown to me, until he returned, back in the CONUS he called my wife in Daytona Beach, Florida and Jack and his wife took Pat out to dinner. Mid-tour was two-weeks and any travel time came off your leave time. I thought giving up a half day of his time with his family was above and beyond. I think that was typical of Jack.

5th ADVON DUTIES

My job with 5th ADVON, was to write Frag Orders (a "fragmentary" order based on an applicable Operations Order) to task the local fighter unit to fly sorties. I said 5th ADVON was a one-man operation when I was transferred; by the time I arrived it had morphed into a lash up of eight people: three other officers, two Non-Commissioned Officers, an Airman plus a Lt. Colonel and a Colonel. Fifth Air Force in Japan, kept track of us through a full Colonel we called Super Colonel. We figured he was above the full-bird rank he carried, but had ticked off enough folks who outranked him, that he would probable not get a star.

Understand this, in order to get the job done without too many people looking over our shoulders, that Operations Order I mentioned did not exist. It was a figment of Super Colonel's mind. If you don't have to coordinate with a plethora of staffers, including ROK units, you don't have to answer a batch of questions. It followed one of my own mantras: If you can't afford a No answer…don't ask the question.

One day, my boss asked to see the Ops Order we used as authority for our frag orders. I told him it did not exist. He turned several shades of crimson and I feared some blood vessels in his head would pop. He said, "We can't operate like this; we have to put the Ops Order in writing."

I said that wasn't a good idea. It would take an inordinate amount of time and in the end be a useless document of no importance. It did, it was and I was right. The final product was hardly worth the paper it was written on, but it satisfied our erstwhile Lt. Colonel and Colonel. Needless to say, Super Colonel was not pleased with our product.

Super Colonel made frequent trips to Korea to see what we were up to and check with the F-4 Phantom II (a two-engine jet) fighter squadron. When on scene, he took over our Colonel's office. On one such trip, he emerged from the office and announced he wanted two Phantoms in the air loaded with air-to-air munitions. He went back into his office leaving us to write the Frag Order specifying our now written Ops Order as authority.

We had the order nearly complete when someone asked how to call for two airplanes, that is what is the plural of F-4. They wanted it to read "two F-4" but didn't know how to make it plural. The argument lasted a good half hour. I suggested using standard terminology—2 x F4—the "x" standing for the word by. I didn't come up with anything fancy, that terminology was reasonably standard in frag orders. The argument continued vacillating between F-4's and F-4s as alternatives. I even made a trip across the street to the headquarters building to check with one of the civilian secretaries and her copy of "The Secretary's Handbook."

The *discussion* was still ongoing when Super Colonel re-emerged and asked if the Frag Order had been sent. He was disgusted with our response and picked up a phone. He dialed and when the other end answered he addressed the F-4 squadron commander by his first name. He also said: Give me a pair in the air, loaded air-to-air. In less than five minutes we could hear four afterburners light off as the two fighter jets roared down the runway.

LIVING IN THE LAP OF LUXURY

I lived up a hill from the main base in a two-story BOQ which while modern, was a pre-fab modular building. As I recall, the only stairs to the second floor were on the exterior of the building. The main room was about twelve feet by twenty-four feet with a large two-door chifforobe in the middle simulating two rooms—bedroom and living room. My bathroom was on the far side of the living room. Also see Figure 91 on page 153.

One would think a person could make it about thirty feet to the bathroom in the middle of the night. Apparently not; after a bout with a bottle of Cold Duck at a party on base, I felt the urge and rolled out of bed. I did make it to the bathroom door, but the Cold Duck insisted on coming up at that point. Nothing like puking purple.

Speaking of the Devil Rum, the Officers Club offered Happy Hour seven days a week.

During Happy Hour, call whiskey, my favorite of the day was Canadian Club, was fifteen cents. When Happy Hour closed, the drinks went up to a quarter. Early in my tour, a gent at the bar asked me if I wanted to roll dice for drinks. That usually consisted of using six dice in a dice cup for a game of Horses. I don't remember the rules, but it took several rolls to determine a winner. I asked if the game would be Horses and here was his answer.

"No! Korean Zap – one flop, nothing wild. Did you come here to drink or roll dice?"

TRIPS OUT OF COUNTRY

Due to my clearances above Top Secret, I was one of a few qualified to attend monthly conferences in Yokosuka, Japan. Yokosuka was a navy base, and while it was a chance to get outside South Korea, it wasn't much of a treat. With that in mind, at one of the meetings we set the next conference date and base. We chose John Hay Air Base in the Philippines, which was known as a pearl of a location. I didn't get to go to that next meeting. My boss did.

When my time for mid-tour leave came, I pulled a few strings. As I mentioned before, travel time came out of the two-weeks allotted. Hooking a ride on a bird out of Osan Air Base was a near nil possibility. I figured if I could get to Kadena, Okinawa my chances improved. I asked my boss about sending me down there to coordinate with an office we worked with. He said no; maybe if they had asked for me it would be different.

He left me an opening. I called the office in Kadena, talked with a friend who worked there and explained the situation. Lo, and behold. The next day a message arrived in our office asking for me to drop in on them and *coordinate* things. Actually, this was only one of several speedbumps my boss tossed out—and I countered all of them. My boss relented and I got a seat on a plane headed for Kadena. First hurtle down.

I arrived at Kadena and found GIs, who had been there a week waiting for a hop out. I signed up for a flight to Hawaii, where I would meet my wife. She flew out of Florida, got to Los Angeles airport and was told her flight was overbooked and she had no seat on the airplane. Then the counter person said, if you can make it to another gate, we'll put you on a Boeing 747 in first class. Pat said, "Which way" and was soon ensconced in first class sipping champaign.

Meanwhile, back at Kadena I was hanging around Base Ops weighing my options, which were between slim and none. Then, the sun came up. I looked around and saw Big Dan, the Tanker Man walk into the room. Same guy I went through KC-135 upgrade training with. I told him I was trying to get to Hickam Air Force Base, Hawaii. Dan told me he was flying C-141s and was on his way west to drop a cargo load in-country Vietnam. If I was still there tomorrow morning, he'd hold a spot on his plane for me when they got to Kadena on the return trip.

I was still there, and got on Dan's bird. Not exactly a "seat," more like space on a pile of soft baggage just behind the cockpit. I didn't care because I made it to Hawaii in two days. Might not be a record, but way above average.

I was waiting by the gate when Pat came down the jet-way showing the effects of first class champaign. We stayed in a small motel near Waikiki beach, sipped Mai Tai's at Fort Derussy on the beach and at the Hickam Officers Club watching ships move up the channel to Pearl Harbor.

[92] Pat arriving in Hawaii

Pat (left) and I (below) am receiving a welcome to Hawaii and getting a

[93] Author greeted in Hawaii

traditional flower lei.

[94] Pat in sunglasses

Sightseeing around the island, I snapped two pics that that are among my favorites of Pat.

Figure 94 and this one (95) are the favorites I mentioned.

[95] Pat on the beach

[95a] Pat on a lanai

The lanai in Figure 95a is most likely Fort Derussy – an old army post on very expensive property; Waikiki Beach.

Not sure what prompted the author to be a tree climber. Note the Bermuda shorts, right is style. Probably purchased on Johnston Island.

[95b] Author climbing trees

One night a second-story burglar decided to try our room. Yes, we were on the second story. I awoke to his scratching at the glass jalousie window panes. I looked for a weapon. The best I came up with was a pocket knife with a two-inch blade. Armed to the teeth, I approached the window with not plan in mind. I decided my best chance was to scare him off before he came into our room. Knife in hand, I shouted and cussed at the top of my lungs. My non-plan worked. I heard him scramble back down to the ground.

Next, I wanted to call the police. Did I mention our motel was in the economy class and did not include an in-room phone? I locked the door, went down to the lobby, called the cops and returned to our room. The police made out a report, since the evidence was slim, they didn't hold out much hope of catching the dude, but did say it was unlikely he'd return.

WHO'S RUNNING WHAT?

The TACC (Tactical Air Command Center) was a tiered facility with large, clear plastic plotting boards at the bottom and was split down the middle with the USAF contingent on the left (looking down to the boards) and the ROKAF members on the right. The plotters behind the plastic boards were Korean.

Technically, the Ops Order spelling out how the TACC would operate said it was a joint effort, but the USAF was in charge. One night, the plotters began putting information on the board in Korean. The senior USAF officer leaned over to his Korean counterpart asking what was going on. The ROKAF officer told him it wasn't important and would explain in a minute.

Unknown to the USAF side of the TACC, the ROKAF discovered North Korean skunk boats (fairly small fast attack boats) heading south of Kangnung in the northeast part of the country. These boats were often used to drop off foreign agents on the shores of South Korea. Without telling or informing the USAF guys, the ROKAF launched several of their F-5 attack fighters (single seat, modified T-38 supersonic USAF training planes) and they were on target and sank the skunk boats before anyone on the USAF side had an inkling of what was occurring.

I suppose the Koreans honored the Ops Order and let USAF be in charge ... sometimes.

MY TURN FOR THE S- - - - Y END OF THE STICK

Got to lead this lash up because I had those high levels of classified clearances. A US Navy reconnaissance bird was scheduled for a tour along the DMZ between North and South Korea. The F-4 fighter squadron would be tasked to provide Combat Air Patrol protection to this plane. By the way, the Navy bird was a slow mover, a prop job that would be making his run at 200 to250 knots. Way slower than the Phantom which was capable of twice the speed of sound.

Word came down from the Air Division Commander's office: the F-4s would be tasked for a co-altitude, co-airspeed profile to match the slow mover. I talked to the F-4 squadron commander; he thought we were nuts; but then so did I. I suggested he put of a pair of birds up and fly a practice profile.

They did. The Phantom has stages of afterburner. At the scheduled altitude and airspeed, the F-4s had to light min-burner on both engines just to stay in the air and burning copious amounts of fuel all the way. Now, imagine a bogey appears and try to figure out how long it will take the escorts to accelerate to 500 knots or better and engage the enemy.

Presented with these facts of life, the previous decision held fast. One of the jocks who flew the practice mission asked me what the idea of protection was. I said the only thing I could think of was to give the bad guys four more engine exhausts to shoot at. He shook his head and agreed.

Along with this problem, I managed another. We had to deliver a frag order and briefing to three other fighter airbases. My boss said, since I had the clearance, I would be the one to make the trip. I'd been logging time in the C-47, yep the venerable old Gooney Bird, so I worked with the Base Ops Officer to schedule the flight to the three bases in a Gooney. Each stop would have a half hour on the ground so I could brief the fighter squadron commanders.

With that much accomplished, my boss scheduled a time we could brief the Air Division Commander on our progress. My first time in the old man's office. Ostentatious was my first impression. The brief went okay until I used the wrong pronoun. As I explained the C-47 round-robin to pass the word to the necessary squadron commanders, I said, "To get this done, I've laid on a C-47 flight to…"

He said, "I'm the only one on this base with authority to schedule flights."

Remember, I did this at the direction of my boss. Remember, my boss is sitting in the room. Could one expect to have his boss take some of this heat. Nope, absolute silence from him. I could have dumped it back in my boss's lap, but I never operated that way. I remained silent on that aspect. Instead, I continued the briefing with the proviso "with your concurrence, Sir, I will …"

The brief came to an end with the Division Commander commenting that I was the worst staff officer he'd ever seen.

Ya take the lumps and kudos as they come. I later heard the General went from Korea to a plush assignment in Europe and came home before the tour was up. Maybe just a rumor, but I smiled and still carry his last words as one of the invisible "medals" I wear along with the other awards I earned.

I know what you're thinking; what happened to the Navy bird? I don't know for sure, but I think it was a case of the crews getting the job done despite the staff. The mission went off without a hitch. Nothing to base this on but conjecture, but I'd guess a pair of F-4s did a co-altitude, co-airspeed escort. I'd also guess another pair of Phantoms were hanging around on high CAP with airspeeds around Mach 1. Maybe another of those cases where if you cannot afford a No answer, don't ask the question.

THE WORD CAME DOWN

Higher headquarters notified us we would be hosting a joint exercise involving U.S. and ROK units. It was to basically be a paper exercise, i.e., we would use forces that existed strictly in the minds of the players.

Our first task was to design an operation that did not mimic the way we ran the Vietnam air war. That country was divided into six Route Pacs (packages) and only one service was responsible for a Route Pac. That made it easy on the staff and allowed tasking of air assets without coordinating with anyone else. That caused more problems than I have time to recount. Back in the day, a reel-to-reel tape circulated with fighter songs of the day and a faux interview between a captain and a newsman with a guy from Public Relations trying his best to get the story straight in the best possible light. All the way through the interview, the captain keeps bad mouthing the way the war was being fought. In the end, the PR representative agrees and says, "What the captain means is, this is a f—-ed up war."

To avoid that, we devised what later was dubbed the JAOCK – Joint Air Operations Center, Korea. The 5th ADVON team imagined eight to ten schedulers around a table with a couple of horse holders for each to act as the gophers. We ended up with a pair of S-80 tents which as I recall were about 20 feet x 30 feet each. Also on paper, the exercise would start several weeks before the active participants step into the arena.

The scheduler's objective was to sort out targets, apply the best asset for a given target—no matter the service owning the asset—and coordinate any conflicts in the air for planes heading for other objectives.

FIRST GLITCH ARISES

On top of the weeks long build up, which only happened on paper, none of the message traffic was supposed to leave Korea. Even when we were "tasking" units in Japan, etc., they should never see the actual order. We also included a heading like, EXERCISE TRAFFIC ONLY.

At the beginning of the sit-down part of the exercise, got a frantic phone call from the Squadron Operations Officer of an EC-121 outfit in Japan. I knew this officer, probably met him at Yokota AB for the recon meetings. That bird was the Air Force version of the three-tail, Lockheed Super Constellation passenger plane. That Ops Officer had three assets assigned to his squadron and he had a piece of paper in his hand tasking him to cover an orbit off the coast of South Korea 24/7 – twenty-four hours a day for seven days a week. An impossible task; you can't fly three airplanes on that schedule for more than a day or two.

I told him to calm down and I'd contact him on a secure phone. When I got him on the line, I explained this was an exercise and that during the paper lead up to the exercise we had augmented his squadron with a half-dozen extra airframes.

I asked if the message contained the EXERCISE TRAFFIC ONLY line. He said, I read who sent it, that is was addressed to our unit and skipped down to the tasking section. That's when I nearly crapped my drawers. I told him to burn the "classified" document, don't sweat it and ignore anything else that sneaks its way outside the peninsula. I next had a long talk with our central message center.

My part of the program was to assist in coordinating TAC Recce planes assigned to us—on paper. I had two assistants, a pilot and back seater with a *real* reconnaissance background. We were scheduled to run all night and around midnight, I told these two recce jocks they could handle it overnight. They started to object when I told them my background was a two-week course at Myrtle Beach and they lived and breathed the subject. I gave them a contact number and instructions to call me if they couldn't handle something that came up. I slept well and they did the job.

A BRIGHT LIGHT IN THE EAST

For all the hard work that this exercise entailed, I later found a bright side to it. I read a good deal how the military handled the coordination of air assets during Operation Iraqi Freedom. I think someone on that staff had a copy of our after-action report. Seems to me they followed the model we set with the JAOCK.

Maybe I wasn't the worst staff officer the general ever saw.

SHORT TIMER

One last view of this from Jack Donovan. He was known to refer to daily fighter missions bombing North Vietnamese targets: We go into the target each day at 10 a.m. and fifteen minutes later a recce bird comes over the target to take pictures. We go back at 2 p.m. the same day over the same target and fifteen minutes later a recce bird comes over. And they're waiting for us every time. How the hell did they get that smart?

His mantra was: do something different. That's exactly what Robin Olds did. As an F-4 wing commander he loaded his birds with air-to-air missiles, gave them call signs mimicking the F-105 Thud squadron calls, and flew a profile similar to the Thuds. The MiG's came up to force the bombers to jettison their bombs. Those F-4s killed seven enemy planes in a fairly short engagement that day. Do something different. another life lesson I learned.

I finally reached "99 Days To Go" — down in double digits—on my tour. At the O-Club I told the bartender to ring the bell and I was buying the round. A half hour later, after they were inviting people in off the street for freebies, I called a halt to the proceedings. I was smart enough to ring the bell at happy hour and my total bar bill was around twelve bucks.

ALL'S WELL THAT ENDS

With the end of my tour on the horizon, I had job feelers out to 5th Air Force Headquarters in Japan. Air Force personnel came back with — no can do; that would be three overseas tours in a row and the limit is two. Alaska was just another duty station when I went there, now it's overseas?

The scheduled thirteen-month tour was generally cut short and we rotated back to the states a month early. My boss told me they considered me essential and planned to hold me over for that thirteenth month. Ain't it grand being essential; that meant I would spend two Christmases in Frozen Chosen. On my first December there, I got a card from Pat that said "Bah Humbug" which resided on top of my armoire for a full year.

I think my attitude was showing, because my boss relented and I was schedule to depart in early December.

I caught my *freedom bird* from Seoul, South Korea and made my way back to the CONUS and sunny Florida. I had orders in hand that would take me to the 55th Strategic Reconnaissance Wing (SRW) to work for my old Ops Officer from Alaska as a pilot in the recon squadron.

When I arrived in Florida, my wife and daughters took me shopping. All my clothes were *gauche* and *out dated*. I came home with a couple of pairs of bell bottom slacks, similar shirts and wide belts. My family told me I was not only back home but also I was now back in style.

While on leave, I received a call from my Alaska navigator. He was a full colonel and assigned to the Strategic Air Command's Strategic Reconnaissance Center (SRC). Since I was already headed for Offutt Air Force Base, he asked me if I'd be interested in changing the assignment to SRC instead of the 55th SRW.

I made a quick cost-benefit analysis and I think I gave him a yes on that same phone call. I later learned that my old Ops Officer from Alaska was really pissed; he had me in mind in a floating position to work on all manner of flying duties. Oh, well, you never know.

V. SRC (Strategic Reconnaissance Center) & JRC (Joint Reconnaissance Center) The Pentagon

25) January 1970 – March, 1974 Offutt Air Force Base, Nebraska

January 1970 - April 1971, 305 Chateau Drive, Apartment #4, Bellevue, Nebraska 68005

Another of my Alaska navigators was our sponsor at Offutt. We moved into the Chateau Apartments in Bellevue, Nebraska (which is the actual home of SAC, not Omaha). On-base housing was available in a few weeks and we were settled.

BACK IN THE WORLD

We qualified for a four-bedroom Capehart unit on the base. Our unit backed up to the base golf course and our son, Mark, began collecting errant golf balls in our back yard.

The Strategic Reconnaissance Center was buried in the basement of H Wing (the east side of the old original SAC Headquarters building).

My first position in the Strategic Reconnaissance Center was as a Watch Officer in the Control Division. That was a glorified title which meant we worked three, eight-hour shifts around the clock; twenty-four/seven. We kept watch on SAC reconnaissance aircraft flying all over the world. Planes from RC-135s to U-2s to SR-71s to C-130s and Ryan drones from bases from the CONUS to Alaska, to the UK, to Southeast Asia.

The flights were displayed on three by four-foot maps overlayed with clear plastic where the routes were shown in a variety of colored tape. Our job was to keep the info box on the map current so the SRC chief could brief the Director Operations and the CinCSAC (Commander in Chief of SAC) each morning around seven a.m.

One map for each recon mission was hung on a fiberboard backing so the whole batch could be carried from the SRC to the "war room" which was under the front lawn of the building. There could be a dozen or so missions represented on the board and it was heavy.

Those maps had to be in a certain order. If anything changed during the night, the Watch Officer was responsible for the correct order of the maps. The first thing my "trainer" did was to show me the fastest way to pull maps off the fiberboard backing and rearrange the order. May sound mundane, but you didn't want your Colonel to be embarrassed in front of a three-star general.

COMMUNICATING WITH FOLKS

Our area was at least thirty feet deep and one hundred-fifty feet long. Since the entire vault was a classified area, we had special privileges…didn't have to lock up classified papers in secure safes and there were secured teletype machines and a secure phone which sat atop a safe around eighteen square foot monster. Those phones had several levels of use: standard or personal; routine (military related, but of minor importance); immediate (military traffic with a bit of priority); ops immediate (military information that needed to be sent out); flash (top priority, mostly reserved for general/flag officer level) and Flash Override (self-explanatory).

Late one night, a recce bird went off the runway on landing. My first thought was that JRC in the Pentagon needed to know, ASAP. I grabbed the secure phone and the operator asked me who I wanted to talk too. I said JRC, ops immediate. He said the line was tied up on higher priority.

Making tons of calls late at night, we got to know the operators handling the secure phone calls. After a couple of additional denials, I said, I'm not authorized to use Flash, but if I don't get through on this try, I'll go for it.

I don't know exactly what he did, but I was talking the JRC in a few seconds. I found out later that while I was on the phone, an Admiral at SAC asked to be connected to someone and used a Flash priority. The operator said "Sorry, Sir, the line is working equal or higher priority." It was me with my Ops Immediate call.

The Admiral raised hell, and by the end of the day the Ops Security folks were in our office installing a second secure phone. Whatever gets the job done.

MOVING ON UP

Within a few months I was promoted to OIC (officer in charge) of the ELINT Branch. ELINT was shorthand for Electronic Intelligence. We were charged with tasking airplanes around the world which collected electronic signals from the other guys.

I put in several months in this position when I moved up again. I was in charge of the Control Division; back where I started but in charge of the Watch Team. Remember those heavy boards full of maps I mentioned earlier? I know they were heavy because as head of the Control Division, I got to lug it over there a few times when the old man didn't feel like carrying it. Actually, he lugged the board on his own most of the time.

LONG GONE, BUT NOT COMPLETELY

Earlier, I related my trip to the South Pacific with the Rivet Ball aircraft to fly against the French nuclear tests. That operation was called Burning Light. The "word" came down that the French were giving up their atomic tests and all copies of the Ops Order were to be destroyed. Did I mention before that I tend to be a hoarder? I rounded up the copies of Burning Light and consigned them to burn bags…all except one which I squirreled away in a classified safe.

I don't know what gremlin told me to do that, but I thanked him a few months later. The "word" came down again that the French tests were back on! Plan writers were moaning about having to rewrite the whole damn plan over again. It told them I could get the job done in hours and headed for the safe containing the copy I kept.

It took us a couple of days to change a few parts, but we had the Ops Order ready and approved in less than a week. That was only the beginning of my headaches. The SRC boss put me in charge of coordinating meetings in advance of the plan.

If you've ever put plans or messages together, you know there a ton of chops needed before the job is finished and sprinkled with holy water. A chop is a coordination procedure where signatures or initials are required from all interested and even non-interested parties. SAC would maintain overall command and control over the project, but it involved several other departments. As I recall, another command in the USAF (Air Force Communications System) would be there, the Atomic Energy Commission (AEC) from Los Alamos, New Mexico, several other small groups plus the U.S. Navy would have a plane involved.

Timing dictated that all interested parties needed a meeting. My boss said it would be in our SRC briefing room in less than ten days. I got on the phone and began the preliminaries by calling everyone involved. I knew several of the people on the call list so that smoothed the ruts a bit. I got verbal chops from all the agencies as to location and date. Now it was time to formalize the meeting.

I wrote the message draft and personally made the rounds of SAC headquarters. Since outside agencies, Air Force commands, another military service and a civilian agency were involved, I needed an okay from the SAC Protocol office. I met the officer in charge and gave him a quick and dirty overview. When I mentioned the short meeting date, he said that was not possible. Meetings involving other commands and outside agencies required a minimum of thirty days' notice.

I told him the meeting was laid on and firm, and he had ten days to get done whatever he had to do. He didn't like it, but signed off. The fact that I was representing people way up the food chain and above our pay grade no doubt helped.

The meeting went well, the crews got the job done and it was over. Unfortunately, the RC-135 from the Air Force Communications System command was lost over the Pacific. As with many accidents where the plane goes down in the vast expanse of an ocean, nothing was ever recovered and the actual reason for the loss is not known.

I seem to recall a trip to New Mexico to coordinate with the AEC for the after-action report. As we wrapped up our meeting and the report they asked if there was anything else they could do for me. Almost in jest, I said I could use a Joint Service Commendation Medal. They made a note of my comment and I returned to Offutt AFB and went back to work.

I was pleasantly surprised a couple of months later when I was called to the SRC briefing room and the Chief of SRC, Colonel Clancy presented the Joint Service Commendation Medal to me.

Here's a humorous aside. Each morning, before the briefing to the Director of Operations and CinC SAC, there was a pre-brief in the SRC. In addition to folks from the Recon Center, the SAC Director of Maintenance attended to cover any mission glitches which might be charged to maintenance. I mentioned Colonel Clancy before. His first name was Orville. The first name of the full colonel, maintenance representative was Wilber. Not to their faces, but more than once the pair was referred to as the Wright brothers. Oh, well. Seemed funny back then.

WHAT DID I DO FOR FLYING PAY?

With my staff job at Offutt AFB, I was again consigned to Base Operations to fly for flight pay. There I was introduced to the first Convair aircraft. I flew a couple of models, the T-29 which was a navigator training plane with a half dozen or so bubbles on the top center of the fuselage so multiple navigators could take celestial sextant shots at the same time, and a C-131 which was basically the same plane but served as a transport plane. The VC-131 was the same but spiffed up to haul passengers. These were twin piston engine planes with reversible propellors.

We flew a variety of missions; round robins into several Air Force bases to drop off or pick up small priority items, and out and backs to Andrews AFB in the Washington DC area. Occasionally we provided transport to Andrews for the wife of a Vietnam MIA. She was riding and pushing the DC pols to get their act together and provide information on MIAs. I always thought it was an honor to give that lady support. Most of the time, on the round robin flights, we had rated navigators and Electronic Warfare Officers riding along to qualify for their flight pay.

Never understood why the staff officers had the "fly four hours a month or lose your flight pay," especially for the EWOs and Nav's. They just sat on their tokus for the duration because there weren't any airborne duties for them. They could have been sitting in their assigned office being productive at the job they were assigned to. USAF policy wonks never asked me for an opinion.

On one of these flights, Al (Lurch) Hansen dropped that one-liner on me. I included that story previously.

I was returning from the east coast in the VC-131 with an Admiral aboard on a nasty night flight. Normally we didn't have to refuel at Andrews because we had enough fuel for the out and back. This night, we were dodging bad weather and I was concerned fuel would be tight for getting back to Offutt. Rather than chance it, I re-filed in the air for a refueling stop at Wright-Patterson AFB in Ohio.

On the approach, I notified traffic control of the number of people on board and mentioned honors letting the base know we had a Flag Officer on the plane. That meant the Base Commander or Wing Commander would most likely meet the plane out of courtesy. A full colonel was waiting at our parking spot and welcomed the Admiral who said he didn't need anything special. The humorous part was that I knew the Colonel from a previous assignment and he was wearing his "fancy duds." That was a custom-tailored flight suit with patches and gee-gaws embroidered all over. The Admiral got a chuckle over the clothes, we refueled and were on our way, shortly.

Here's a Navy aside. At some point when I had an aircrew flying the KC-135, we were diverted into a Naval Air Station for some reason—which escapes my mind at this point. Approaching the NAS, we made the normal call to let them know how many SOBs (souls on board) there were. They asked for honors, and without thinking where I was, I said we had three captains. I parked the bird, looked out my window and saw three, count them, three staff cars waiting for us.

All three of us were captains, Air Force captain which are 0-3s; the Navy folks assumed I was talking about 0-6s which is the Navy equivalent of a full colonel. I thanked them, released two of the staff cars and we piled into a single ride. I'm sure someone on the base had a comment or two about those dumbass Air Force people.

WHERE ELSE CAN YOU HAVE SO MUCH FUN?

On this T-29 flight, we were headed to Peterson Field near Colorado Springs, Colorado. It was a great day, clear and bright and approach control aimed us at a 7 to 8,000-foot-long runway. An old friend, a neighbor from Alaska was flying right seat with me that day. On fairly short final, the controller said they had runway problems and would I accept a different runway. The heading was only a thirty degree or so offset so I said we would accommodate their request.

My friend in the right seat checked the particulars for that runway, smiled at me and said, "That's a 2,000-foot-long runway, John." As I said, the weather was great, little wind, and we had prop reversers or probably would have aborted the flight. With a touchdown at the end of the runway and those paddle blades running backwards we had no problem stopping.

Indulge me; just one more T-29 story. A typical round robin to hit several bases to drop off items. The weather was not cooperating that day, and our first scheduled stop was Whiteman AFB, just east of Kansas City, Missouri, was below minimums—ceiling too low for landing. The rules say you cannot file (to land) for a base below minimums. I filed Whiteman as a waypoint and then onto our next stop which had decent weather.

I briefed Base Ops that I would overfly Whiteman, and if the cloud base was above minimums, I would refile my flight plan in the air and make our scheduled stop. Seemed like a plan to me. It worked too. When I arrived in the Whiteman area, the ceilings were rising, I executed my plan and landed. Right here and now you have to understand that the IG (Inspector General) and ORI (Operational Readiness Inspection) teams loved to sneak into a base unannounced. And, they often used T-29s to do the dirty deed.

So, here was an "unannounced" T-29 parking on their ramp. I shut down one engine and we prepared to offload whatever valuable cargo there was for this base when the Fit hit the Shan. The tower was calling me on the radio to let me know that the Base Commander was hollering that he wanted to see the pilot of this unannounced T-29 inside Base Ops—ASAP. That means to report quickly and smartly in a military manner RIGHT NOW.

I asked the tower to convey my apologies to the Base Commander, that Offutt Base Ops must have dropped the ball as to my intentions and that I was not planning on shutting down engines. In fact, I was about to restart an engine and get moving out to complete the rest of my flight for the day.

I'm sure there was a full-bull sputtering around Base Ops when I taxied out for takeoff.

Don't-cha just love it when a plan comes together? I never heard word-one about our passenger stop at Whiteman that day.

PROMOTION TIME

The promotion list for O-5, Lt. Colonel, was due out shortly. I contacted my Alaska nav who was now in the Pentagon. When he called back, all he said was, two, oh, five. That told me I was on the list for promotion and my line number was 205. Early release of all or part of the list was frowned on; but that didn't' stop many.

The promotion list details were made public. The first two-hundred on the list would be promoted on the release date. I wouldn't be in the first batch, but the second group should be out in a month after the first ones.

Not so fast, Hoss. Tricky Dick Nixon got pissed at the military for some reason and he put a freeze on all promotions. We all sweated out our promotions and about sixty to ninety days the unfreeze order came out. Eventually, the pay raise we missed was paid.

[96] Author on motorcycle

While still living on base, I decided to add another mode of transportation to our household. This is a Honda 100—it was actually about 90 cc's instead of 100. I rode it for about a year and a half and it never put me on the ground. It was built like a dirt bike, but carried enough equipment that it was street legal.

Heeding the adage that there are two types of riders, them that's still upright and them that's laid one down; I sold it.

May 1973 - April 1974, 707 Sherman Drive, Bellevue, Nebraska 68005

We purchased a home in Bellevue and the move was smooth. I remember signing a thirty-year mortgage and almost panicked. After all, this was the first home we bought in our lives. It was a great home, and our children didn't face a move for some time. This picture shows the house and she's in darn good shape considering I took this photo in 2017.

[96] Sherman Drive 2017

I was nearing a point I knew SAC would be looking at me for a transfer. I put out feelers about an on-base transfer to the JSTPS (Joint Strategic Target Planning Staff), which was in the headquarters building, the same building where I went to work every day.

Didn't start early enough. Personnel sent me notice; my transfer was imminent but not here at Offutt. In their infinite wisdom they chose the Pentagon for me.

The general in charge of the Joint Reconnaissance Center (JRC) in the Pentagon fired a Watch Officer on the spot leaving them shorthanded. They scattered the IBM cards on the floor, stomped all over them with their golf cleats and of course my card came to the top. Ripe for transfer, all the right security clearances and still breathing.

In their panic to cover some dingbats fit of pique, I only had a little over a month until I had to report to DC. In the Air Force there is a "sixty days orders in hand" rule; meaning I was supposed to have two months to report to my next duty station. I checked with personnel and a young airman told me I would have to sign a waiver before they could cut my orders. I would have to waive my rights to cover some general's ass.

When I told him that I would report on time, but I would not sign their stinking waiver, he said he'd have to send a message to the Pentagon. I said, "Son, if that's your drill, you'd best get to it."

Next, I received a call from my old Alaska navigator again. He had left Offutt and was now in the JRC. I repeated to him that I would be there on time and they could stuff the waiver. Never saw so many upset over a piece of paper. I figured if I signed the damn thing I got them—whoever them was—off the hook. Now they had to justify transferring me on short notice.

A WORD ABOUT SAC

Before I depart for Washington D.C., I need to say a word or two about the Strategic Air Command (SAC). I would bet that every member of SAC, cussed that damn command at one time or another during their time in SAC. That said, SAC was one of the best, if not the best command in the USAF.

It was formed in 1946 as WW II wound down. Within two years, Curtis E. LeMay took command of SAC. He was already successful with the strategic bombing plan in the Pacific Theater, and he held the helm from 1948 to 1957. For all his faults, he was a hell of a commander. Here are a few quotes he left behind.

"We should always avoid armed conflict. But if you get in it, get in with both feet and get out as soon as possible."

"War is never cost-effective. People are killed. To them, the war is total."

"Successful offense brings victory. Successful defense can now only lessen defeat."

"If I see that the Russians are amassing their planes for an attack, I'm going to knock the shit out of them before they take off the ground."

"The price of failure might be paid with national survival."

SAC was a centralized command, i.e., each wing and squadron commander had the responsibility and authority to manage his (weren't any she's back in the day — our loss) men and equipment as he saw fit. However, the orders came from the men burrowing under the front lawn at SAC headquarters at Offutt AFB, Nebraska in the bomb "proof" underground command post.

I viewed SAC as a Combat Crew member, and then as a staff officer in the Strategic Reconnaissance Center (SRC). At SRC, we received tasking directions from higher headquarters and put the anticipated flights into a package. That package went up to the Pentagon and then to the White House. When everyone gave the package a thumbs up, SRC parceled the individual flight plans and targets out to recon units around the world.

At the squadron and wing level, each mission was monitored in detail from maintenance to preflight, to takeoff, through the mission to landing and debriefing. In the SRC, we were notified of the plane's takeoff, and specified reports during the flight and the landing time. At each event, SRC received the information via secure teletype machines which linked the units to SRC. Typically, we didn't bother the Joint Reconnaisane Center (JRC) in the Pentagon unless something went amiss.

The reports I mentioned earlier, were simple "Ops Normal" messages sent over and HF (high frequency) radio network. Ops Normal says everything is okay. Since they were simple and the format didn't change, each command post had a stack of "canned" message formats and when a bird reported in, they only had to enter a couple of entries and hand it to the folks who transmitted them to us.

Late one night when I was on duty as a Watch Officer, a unit sent us a typical ops normal message about the mission in progress. Minor problem, the guy on duty grabbed a blank message form and gave it to the message center. All I received was the canned format with absolutely no information. I could almost see the fellow on the other end and thought we'd have some fun. I called them on a secure phone line and said, MESSAGE RECEIVED, LONG ON FORMAT, SHORT ON INFORMATION. We both laughed and I got the correct message in a couple of minutes.

Any SAC crew dogs out there will no doubt remember transmitting those Ops Normal reports and listening for the HF radio to blast out a message that began, "Sky King, Sky King…this is Dropkick with…"

MORE ABOUT SECURE COMMS

In the SRC vault, probably 2,000 square feet, we also had a secure phone which could reach around the world. The use was controlled by the priority assigned to each call. There was the lowest, basically personal available to anyone with access to the phone could use it as long as no one with a higher priority wanted it. It was a one call at a time system. Beyond that one was Routine (real business but at the low end of the scale), Operational (related to routine info about flights), Ops Immediate (Operational Immediate – about flights and things were heating up), Flash (major important information) and Flash Override (the Fits about to hit the Shan). (Also mentioned earlier.)

In my job, I was limited to Ops Immediate, which was enough to get the job done most of the time. Flash and higher was generally limited to General and Flag officers. To put a call through, we picked up the receiver, which sat on top of a three foot cube, and waited for the operator to come on the line. Keep in mind this was not the stuff you see on TV today…it was primitive and slow. Hanging on the line waiting for a call to go through we got to know the operators voices and they knew us.

A BIT OF NON AIR FORCE FLYING

Somewhere between Florida and Nebraska, I decided to get a flying license from FAA (Federal Aviation Administration). I knew I could pass the written exam and I wouldn't need a flight check; the FAA would accept my Air Force flying time as proof of being able to fling an airplane around in the sky.

I passed the Commercial Pilot's license exam and was duly issued a certificate to fly and charge people for the privilege of going with me. Never did that part. When my license arrived, I looked for the listed type ratings, that is, what kind of airplanes would they allow me to fly. The image here displays them: I was qualified to fly multi-

[97]-Commercial pilot license

engine jets like the Boeing 707 and 727. I could also take up multi-engine prop planes like the Douglas DC-3, and Convair's CV-240 and their sisters. I could also fly these planes in instrument conditions (IFR).

The types of planes were based on my most current time in the Air Force. Since I had not flown single engine airplanes for years, I was not qualified to fly a single engine plane of any type. When I got to Offutt, I joined the Offutt Aero Club and checked out in the Piper PA-28, Cherokee. It was a four-place, single engine prop plane and after a few rides with an instructor, he turned me loose.

I was going to solo again. I took off, stayed in the traffic pattern and turned downwind. I was cruising along looking at the scenery and the runway when it dawned on me—Good Grief! I had not been alone in an airplane in more than a decade. Guess that's why they call it solo. I didn't have too many hours in that Piper, but it was enjoyable flying nowhere with no limits on the flight. It was pretty reasonable by those times, around $25 an hour, wet. Wet meant the hourly charge included fuel.

Here is the danger of writing a Stream of Consciousness narrative. Occasionally, the stream runs over the rapids and repeats itself. I included this next story earlier, however it's such a hoot to me, I'll leave it in here for the second time. Your choice: read it again or skip it 😊

Late one night, there was an accident at Okinawa. An SR-71 Blackbird (faster that a speeding bullet which flew at altitudes where few dare to go) ran off the runway at Kadena Air Base on landing. Minimal damage, and no injuries, but it I didn't get the info to the Pentagon immediately they would probably fry our asses. I grabbed the secure phone and when the operator answered I asked for the Joint Reconnaissance Center on Ops Immediate. The operator said, "Sorry sir, it's working a higher priority." That meant I had to wait.

I went through this drill a couple of more times when I said, enough's enough. I told the operator that my authority was limited to Ops Immediate, but I had info I needed to get to the Pentagon and the next time I picked up the phone I would ask for Flash priority. He said, 'Yes, sir, let me see what I can do." The next thing I knew I was talking to the JRC. That done, I hung up and almost forgot the call.

The next day when I returned for my shift in the SRC, I found we had a second secure phone by our console. Being curious, I asked why? Here's the backstory. Last night when I was on the phone, an Admiral here at SAC headquarters lifted his secure phone and asked for the Pentagon on a Flash priority. The operator told him that he was sorry but the line was working equal or higher priority and he would have to wait. He was steamed. So, the next morning the Comm guys showed up and installed that second phone.

Apparently, no one asked any detailed questions. I never heard a word about my "equal or higher" phone call.

HOW MUCH TIME IN SAC?

I served nearly three-quarters of my career in SAC—around fourteen years by my calculations. For all the cussing and having to enforce the "despite the staff, the crews will get the job done" edict, the Strategic Air Command was a damn fine outfit to serve with, however…let me poke a bit more fun at them.

I'll use one last quote attributed to Curt LeMay, "To Err Is Human, To Forgive Is Not SAC Policy," and combine it with a comment by Charles Colson (tricky Dick Nixon's sidekick) when he said, "If you've got them by the balls, their hearts and minds will follow," and link them to the SAC crest.

[98] Original SAC crest

The diagonal banner is the Milky Way. The SAC crest shows a mailed fist holding lightning bolts and olive branches. The Motto was: Peace Is Our Profession.

Based on Colson's motto, then here is my version …

If you can't make fun of yourself, who then?

[99] SAC crest modified

A little something different.

[100] 24SRS Crew patch

A patch I designed to represent our crew, E 10.

Diagonal letters are our crew number, E-10 with the E in phonetics. Bottom left represents Burning Candy and Combat Apple SIGINT missions we flew and in the upper right is Rivet Ball, PHOTOINT. The black ball is from a pool table and represents Rivet Ball and the 18 SOB.

26) March 1974 – January 1976 Pentagon, Washington, D.C.

MY TRIUMPHANT ARRIVAL

Not! You would have thought I'd been riding on a manure cart when I arrived, ON-TIME. Being on-time didn't buy me any points. I lost them all by refusing to sign that waiver. And I never got to meet the general who set the whole thing up. There was a new boss in the JRC when I arrived.

Our oldest daughter was going into her senior year in high school and was not looking forward to another transfer. Pat and I discussed the situation and we decided I would go ahead, and check out the area before we sold our house and moved. Real estate prices were through the roof in DC; getting to our price range would mean going forty miles or so outside the Beltway. We decided not to move the family; I would commute between DC and Nebraska.

GETTING SETTLED AND TIME OFF

The console was a long bench arrangement where the Watch Officer and an NCO kept an eye on worldwide activities in the recce world. Both positions at the console had a build in phone with fifty or sixty buttons, most of which were unimportant. We soon learned which ones carried important information when they blinked.

At night, the watch team were the only ones in the JRC. We had a large TV in the room that provided respite from long dull nights. No one could enter without our knowledge. They were stopped at our entrance door, had to punch a button to alert us and look up at the CCTV. From the watch stations, we could identify the person and remotely buzz them in.

I located a room rental in a house on Columbia Pike in Arlington, Virginia. It was only a few miles from my work place and since shift change was seven a.m. and p.m., traffic was not a problem. The house was owned a retired (medical) Army officer and Vietnam veteran who build a new house next door. There were five of us unaccompanied folks living there.

At full strength, the watch team would work two twelve-hour day shifts, then two, night shifts followed by six days off. That was at full strength. Want to bet how often we were staffed at that level? That's right, not often. That meant two days, two nights and four days off. Better that nothing.

United Airlines had a direct flight from Washington National Airport to Omaha, and the fare was reasonable. I could check with Andrews AFB, Maryland, to see if there were any Space A seats available on military planes.

A four-day break made it difficult to do a round robin; six-day breaks made it easy. Either way, I got in a lot of sight-seeing in the DC area. Back in the day, there were two sides of parking on both of the streets at the National Mall.

If I couldn't find a spot there, I headed for the Washington Monument. There was a time limit on parking, one hour, I think. If they caught your overtime parking, the fine was five dollars. Paid parking was way more than that, so even if caught it was a bargain. I spent tons of time in the Smithsonian, National Archives, Supreme Court, Library of Congress, Lincoln Monument, Washington Monument and the Capitol.

On one break, I didn't feel like the usual sightseeing venues. I drove to the Iwo Jima Memorial. It was a bright, warm day with a comfortable breeze. I walked to the statue, found a shady spot and sat down on the grass. For entertainment, I spent a half-hour watching a stream of ants moving across the grass and up one of the large trees nearby. It was a rejuvenating interlude with time for contemplating navels and life.

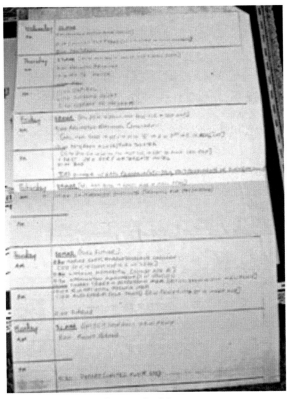

[101] DC trip itinerary

I had to do a lot of planning for this one, but I managed to get a long enough break to bring two of our children, Mark and Karen, to DC for a visit. I put together an itinerary that covered every minute of the time they were there.

This is still BC (before computers) and my itinerary (Figure 101) was rather crude.

They arrived on a Wednesday evening and departed the following Tuesday evening.

We visited many of the sites I listed above and tossed in a trip to Arlington National Cemetery and a White House tour.

I put a scrapbook together for each of them filled with pictures of them visiting memorable edifices.

[102] Scrapbook cover – DC trip

The trip covered the last week in March of 1975, and I used April on the cover text.

A graphic artist was assigned to the JRC. When he got wind of this visit, I was planning for our children, he designed a cover featuring many of the sites we visited. This particular one belongs to Mark, and he sent the two images to use here.

We are lined up for a tour of the White House, author behind camera.

Mark is in the center (red slacks) and the only guest looking at the author and the camera.

Karen is the lady to the left of Mark, the one with all the hair.

[103] Karen & Mark in WH line

Still waiting in line, but much closer to the White House.

[104] Mark entering WH

Mark steps out of line for a better view and the portico columns are at the left of the photo. We're close now.

I wasn't required to fly for pay. Before this assignment, I needed four hours every month or lose my flight pay. In many ways it was a ridiculous requirement. Look at this setup: we have a line Pilot and a line Copilot in the seats logging pilot time. I'm along on this ride and I'm an Aircraft Commander; I can log AC time for the entire duration. To round out the crew, add an Instructor Pilot and he logs IP time. The flight is four hours long and the four of us (P, CP, AC, IP) all log four hours of flying time. That's sixteen hours of time logged on a four-hour flight, and all of us qualify for flight pay this month.

SPACE A (AVAILABLE)

I always checked Andrews AFB to see if any birds were headed to Offutt AFB; much cheaper than commercial. On one break, Andrews told me a major general – a three star—was headed that way in a T-39. The Saberliner was a small, twin jet utility and training plane which could carry about a half-dozen passengers. I boarded the bird and introduced myself to the general. He pointed out his horse holder, a Marine full colonel who sported a typical buzz cut on his hair. The general was seated in the one seat that was opposite a rear facing seat. He looked at me and pointed to that seat. That left a seat on the opposite side of the plane for the Marine. He didn't look happy and scowled at me.

The general and I hit it off. We had some comparable background and, as was I, he was wearing Wellington boots. Not exactly by-the-regs footwear. I was getting toward the end of my tour and my hair was a barely acceptable length. That also matched the general's haircut. It was comfortable ride and the time went quickly. I'd never hobnobbed with a three star before. The Marine was still scowling when we got off the plane at our destination.

I was wandering around Base Operations on another Space-A excursion when I ran into a major, I'd known in Korea. We chatted for a few minutes when he said he needed to get back to his group. I was sporting the same footwear and haircut. My friend returned in a short time. Apparently, his boss had spotted me and was asking questions.

The fellow said, "My colonel wants to know where you work." I said, "You can tell him I work on the Joint Chiefs of Staff at the Pentagon." I guess that satisfied *his colonel* because my friend didn't return with more questions.

ON DUTY DAYS

I looked at the COPS, Console Operating Procedures, and compared to what we had at the SRC watch desk, they were a mess. I volunteered to do some rewrites. Too late, I asked who wrote them. My boss said, "I did." Oops.

Shift changeover was seven a.m. and seven p.m. Twelve hour shifts except for the two Daylight Savings Time changes. I got the thirteen-hour shift while someone else drew the shorter one. At full strength, we did two day shifts, then two night shifts and were off duty for six days. We were seldom at full strength, so it more like two on, two on, four off.

At the evening change over, the day guy brought his replacement up to date. Overnight we kept track of recon flights all over the world. If nothing untoward happened, all we had to do was troop to the NMCC (National Military Command Center) next door and exchange message traffic. All the watch officers had access to that area, but only the top two in the NMCC were authorized in our vault. A General/Flag Officer sat in the "cab," a glass walled office area that overlooked the large area that made up the NMCC. A full colonel or navy captain was his second in command. Out in the main area, folks from all manner of agencies and services sat at desks along with specialists in areas of interest around the world. (If you read my thriller, "Assault on the Presidency," the above section will sound familiar.)

Around seven in the morning, we briefed our replacement so he had enough detail about the night he could brief our boss, the chief of the JRC.

I've mentioned the high cost of living in the DC area. We had three-stripers assigned to us. Back then I recall that rank was an Airman First Class, four rungs up from the bottom. I shook my head wondering how they survived with a family. I recalled a tech from the National Security Agency (NSA) scanning the SRC vault. I asked if they found many bugs, he said it was easier for the bad guys to "buy a body." Those low-ranking folks would be prime targets. I don't remember hearing of anyone who did that. That didn't hold true for long.

BIGGEST DEAL

I've mentioned before that with being assigned to a reconnaissance slot, we were cleared for higher levels of classification. These were above our Top Secret clearance and the names of those levels were classified. The names were never spoken outside a secure environment. One day those names were changed. Seems Secretary of Defense McNamara waltzed out for a press briefing with a folder under his arm. That red bordered folder cover also was emblazoned with the name of one of those classified levels. That's what caused the abrupt change.

As if that wasn't complicated enough, special situations called for a compartmentalized level. First you needed to be cleared for the proper level, and then you had to be "read in" on the project. That occurred in the summer of 1974. The JRC boss, his deputy, the chief of the watch team and the watch officers were the only ones who were read in for the Glomar Explorer Project Azorian (Project Jennifer).

At this point, I have to admit that my memory did not retain a lot of the details we were privy to at the time. As I wrote this, I was forced to do a Google search to find the unclassified names assigned to this episode.

The wreckage of a Russian nuclear-powered submarine, the K-129, had been pinpointed and an effort to recover the U-boat were underway. During several night shifts I could sense the tenseness through the messages. At the start of the operation, a new, teletype machine was installed in the Watch OIC's office. When the bell on the machine chimed, the Watch Officer dropped whatever he was doing and sprinted the twenty feet to grab the paper spitting out from the teletype. Next the message was logged in and the required notifications were made.

I don't know how the ship was built, but it seemed the vertical activity was moving in inches rather than in yards or feet. The crew was lowering a huge claw down from inside the belly of the ship. Imagine one of the coin operated claws we've all seen in stores where you go grappling for a toy. Then magnify that image up to the size of a football field and that was the Glomar Explorer doing its thing. All this activity was covert, and from the outside the ship was just holding position and doing little.

I didn't share any of these details with anyone. Imagine my surprise when I picked up one of those weekly news magazines and there was a picture of the Glomar Explorer. The article included interviews with crew members recounting more detail than you can shake a stick at.

WHERE DO I GO FROM HERE?

With nineteen years of service, I was weighing my options. If I could get back to Offutt AFB, I was considering staying on active duty. Looking back at past experience with Air Force personnel, I didn't hold out much hope of getting my first, second or third choice.

The Air Force requires ninety days notification to retire. Remember the sixty days' notice orders in hand for a transfer rule? This ninety-day requirement was hard and fast. I made my decision and in October 1975 I told the Air Force I would be calling it a career in January of the following year.

Another Air Force rule says you can only retire on the last day of a month. Since I entered service on January 4, 1956, I would be retired on January 31st with twenty years and twenty-seven days of service.

I was a bit tight jawed, for other reasons, the day I bid adieu to the service. The staff threw a small party in the JRC and awarded me a Joint Service Commendation medal. The same award I received for running Burning Light. I damn near said no thanks, but instead I said, "I'll accept the Oak Leaf Cluster (OLC)," indicating this was a second time for this medal.

[105] A career of awards

1	2	3
4	5*	6
7	8	9
10*	11	12*

*THESE THREE AWARDS SHOWN ABOVE HERE ARE NOT DISPLAYED ON THE NEXT PAGE; NO MEDAL EXISTS

1. MERITORIOUS SERVICE AWARD
2. AIR MEDAL –
   (9 TOTAL) AWARD + 8 OAK LEAF CLUSTERS (1-SILVER [5], 3 BRONZE [3]]
   MY EIGHTH AIR MEDAL (SEVENTH OAK LEAF CLUSTER) WAS AWARDED FOR THE PERIOD AUGUST 8, 1968 THROUGH DECEMBER 20, 1968. THE EIGHTH OAK LEAF CLUSTER (NINTH AIR MEDAL) WAS AWARDED FOR A SPECIFIC FLIGHT ON DECEMBER 18, 1968.
3. JOINT SERVICE COMMENDATION MEDAL (2) - AWARD WITH ONE OAK LEAF CLUSTER
4. AIR FORCE COMMENDATION MEDAL WITH ONE OAK LEAF CLUSTER
5. AIR FORCE OUTSTANDING UNIT AWARD
6. AIR FORCE COMBAT READINESS MEDAL (AWARDED FOR FOUR YEARS ON A COMBAT CREW. I MISSED THE OAK LEAF CLUSTER BY A MONTH OR TWO)
7. ARMED FORCES EXPIDITIONARY MEDAL
8. NATIONAL DEFENSE SERVICE MEDAL
9. VIETNAM SERVICE MEDAL WITH TWO BATTLE STARS
10. AIR FORCE LONGEVITY MEDAL (EACH = 4 YEARS) - WITH FOUR OAK LEAF CLUSTERS
11. ARMED FORCES RESERVE MEDAL
12. AIR FORCE SMALL ARMS EXPERT MARKSMAN RIBBON

WINGS AND THINGS

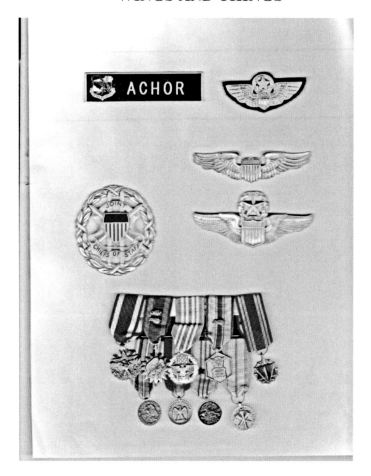

[106] More awards, wings, and badges

Top Left – my Strategic Air Command name badge

Top right – USAF Command Pilot wings (embroidered) for wear with a Mess Dress uniform. I've been told the embroidered type are no longer authorized.

Two wings below (in order of seniority)
    Top – USAF Pilot wings, this is the pair my Basic Instructor pinned on me at Webb AFB, TX.
    Not shown – USAF Senior Pilot wings, like the Command wings without the wreath around the star. Remember the story of 2,000 flying hours and these wings?
    Bottom – my USAF Command Pilot wings.

Circular badge on left – Awarded for a Joint Chiefs to Staff assignment, and can be worn on
    subsequent assignments.

Two rows of miniature medals for wear on a Mess Dress uniform (three of the awards on the previous page are missing here; there are no medals for those awards).

You too could accumulate these awards, and maybe in less than the twenty years and twenty-seven days of active duty it took me. 😊

One positive from these years that is way off topic. I used much of what I lived and learned in DC, Alaska, and my flying career to include in a pair of thrillers I've published. "Assault on the Presidency" and "Assault on Reason" feature a protagonist who is a USAF pilot who flies reconnaissance aircraft and at one point is assigned to the Pentagon. Other than the obvious similarities, everything else in those books are figments of my imagination.

I gave the required three-months' notice of my impending retirement in October, 1975. The military only allows a retirement on the last day of the month. I entered service on January 4th, so I served twenty years and twenty-seven days.

POST MILITARY

27) February 1, 1976 to March 1981 - 707 Sherman Drive, Bellevue, Nebraska

REAL ESTATE

We were still living at the Bellevue address and my wife, Pat, was in real estate sales when I returned to Nebraska. I studied for the real estate exam and passed, only to find I wasn't that good at sales. A fellow I knew was opening a private school, a family business, that offered courses to prepare people who wanted to take the state real estate exam. I discovered I was better at teaching than selling.

As I'm writing this section, a small world aside. Yesterday, I read an obituary in the local daily. It was for the lady real estate broker who hired Pat and me. Sad.

My friend's company was authorized to present our material for students in Iowa as well as Nebraska. I would travel halfway across Iowa two or three times a year to present our material in Des Moines. Some "board" in Iowa called us in to defend our classes saying "you're teaching the test" which was on the edge of legality. If we were using a purloined copy of the state exam, that would have been illegal.

I was called on to give my opinion. I told them I had passed the agent's and the broker's exam in Nebraska. My response also included that anyone with a decent background could design material that would help students answer questions about selling real estate without "teaching the test." That's the last we heard from them.

We set a schedule; I would observe the classes, eight three-hour sessions as I recall, all the way through. Then I would begin presenting the entire course the following time with the boss observing from the back of the room. After that I would take on the classes on my own. When it was my turn to present, the boss approached me after the third session and said, "You don't need me in the back of the room. You've got the rest of the classes on your own. That provided another life-lesson for me.

During some psychological study, I found a stress handler. Close eyes, run through the upcoming scenario imagining the absolute best outcome. Go through the same scenario, but this time imagining the absolute worst outcome. Since I've already faced the worst and best that can happen, the actual outcome will be somewhere in the middle.

So, for that first night on my own, I ran the scenarios and faced the best and worst. I climbed into my car and headed for the class. A couple of miles down the road I had a flat tire and had to walk to a phone booth to call my wife for a ride. Remember this was BCP—before cell phones. Pat picked me up and I told her my best/worst story and was laughing about getting the flat.

I always looked for new and different ways of presenting my information. I even tied to my introduction to a class on the first night. My last name is ACHOR – and I've always told people to pronounce it like a parcel of land. During my intro, I wrote on the board: JOHN 43,560…

I followed the number by this symbol ⊕   Next, I explained the symbol stood for square feet and that 43,560 was the number of square feet in an acre of land. I wrote my last name ACHOR on the board explaining I pronounced it the same but spelled it differently. I also promised that one way or another they would need that number when they took the state real estate exam for a sales license.

I remember a lady in one class, probably in her sixties, looking for another challenge in life. At a break, she bemoaned the fact that the math was scary for her. I gave her permission to think and do math. She accepted my words in the same light hearted and caring manner in which they were given. I made a bet with her; I bet her fifty-cents she would pass the exam. Back in Omaha a few weeks after the real estate exam, I received a letter. It was from that same lady and there was a thank you note and a half-dollar coin enclosed. I won the bet, and more important she passed the exam.

While we did a good job and helped people pass the exam, I could see the writing on the wall. The business was not going to support an employee outside the family.

That was when I began learning about real estate franchises. Real estate sales; were booming and everyone wanted in. the major leagues included Century-21, Re/Max, Realty World, ERA, Better Homes & Garden, Coldwell Banker and Red Carpet to name a few. In the end, a very few lasted; most notably Re/Max, Coldwell Banker and Century 21 are probably the ones with the best longevity.

MORE FRANCHISING

The Realty World franchise came to Nebraska with offices in the Regency area of west Omaha. They were looking for personnel and I applied. In a short time I was brought on board as a BDC (Business Development Consultant) and Trainer. My duties included visiting franchise member brokers and training real estate agents. My area consisted of Nebraska and both Dakotas; a lot of territory and little population—miles and miles of nothing but miles and miles. The total population in those days was around two million: one million in Nebraska and a half million in each of the Dakotas. I learned that Realty World was the first international real estate franchise. It originated in Canada and opened their first U.S. office before Century 21 opened an office in Canada.

We had a "uniform" to wear. Most of the clothing, from sports coats for women and men were navy blue, and slacks and skirts were navy blue or gray. Jackets sported the Realty World logo on the left breast pocket. Uniformity was the name of the game, in clothing and sales training. It was not to stifle individuality, but to provide a standard that everyone could employ. The typical objection I got when teaching how to conduct a listing presentation to sellers, was, "I hate a canned speech." My answer was, "A great deal of thought went into this presentation, and to make sure you cover all the benefits you can offer a seller. If you can write a better one, have at it." I would demonstrate to the class by using our presentation binder and the "canned" words. It was printed in their study books, so they could follow along as I spoke the words with an occasional change of a word here or phrase there when my memory failed. When presented properly, the words did not sound canned and the intent of the words made sense to a seller.

Our marketing folks hop scotched across the territory and the member offices covered a great deal of that geography. I built a schedule to visit each broker every quarter and facilitate a Listing class (five-day course) and a Sales class (a four-day course) every other month. Before visiting a new broker in North Dakota, I phoned him and asked for directions to his office. His response was, "Just take I-29 north to Fargo, and hang a left." Once in Bismarck, his office wasn't tough to locate.

Most of the classroom activity was conducted in our Omaha office, but there were occasional presentations in Scottsbluff, Nebraska. After finishing a week-long class on Friday afternoon, I faced a 400-mile drive back to Omaha. It was interstate all the way, but it took six plus hours of driving. I did a lot of traveling—one year I put 23,000 business miles on my automobile.

Our national conventions aimed to boost the comradery of franchise members and top people and offices were recognized. During my first year on the job, I was awarded the Best Business Development Consultant of the year for the entire franchise. I also honed my skills as a trainer. Realty World, to my mind, had the best training syllabus in the field.

Every new training session brought new challenges. To prepare myself for the upcoming class, on the first morning I would pace from the training room to the front door and back. It became a ritual and our admin staff enjoyed my antics. That was my method of handling any existing anxiety. I always felt that if there wasn't some level of anxiety before a new adventure, I didn't care about what I was doing. At least I wasn't as bad as Helen Hayes, the famed actress, she tossed her cookies before going on stage.

Teaching and training are a great deal like being on the stage. I learned a great trainer had to appeal to all students who learn through other senses. I'm a visual learner; I will look up and right searching for the image that give me the words to express what I want to say. Others, may be aural or tactile learners. Their method of taking in new material may be expressed in their language: I see what you're saying. I dig what you're saying. I hear what you said.

Body language was an important aspect of teaching and selling. It's an interesting field and can be invaluable when assessing situations. The body cannot lie; even if the mouth is. But I had to learn that it was an art not a science. I remember sitting in a broker's office. As I presented my point, he was edging forward in his chair and leaning forward. Both signs of powerful interest. I was right about his interest, but wrong about it being focused on me. I glanced at my watch and it was a quarter past noon, and he was ready to go to lunch. Another, I learned from that.

I began wearing western clothing here in Nebraska, mostly in the form of hats and boots. As I mentioned, our two classes ran eight hours a day for four or five days. I told folks I danced those eight hours for five straight days in high heels, cowboy boots that is. Like the guy said Fred Astaire was a fantastic dancer with all the suave moves. Then his partner, Ginger Rogers, said, "And I do all the same steps, in a long dress, high heels and dancing backwards."

During one of those Realty World national conferences, I met the Regional Director and some of the staff from the South Texas Region in Houston, Texas. A short time later, back in Omaha, I received a phone call from Houston and a job offer. It was too good to pass up.

28) April, 1981 - May 1983, 5800 Lumberdale, #113, Houston, Texas 77092

We put our home on the market, it didn't take long to sell and we headed south to Texas. The population of Houston metropolitan area in those days was around three million. The same as the three-state territory I previously covered. The area was nearly the same because we had offices from Houston east to

Beaumont, south to Brownsville and west to Laredo and Del Rio. The regional offices were on the northwest corner of the loop surrounding the central part of the city. They already had a trainer on staff, so I was brought on board as their BDC.

Our oldest daughter, Kathy, was married in Omaha and our second daughter, Karen, had an excellent job there, so Mark was the only one to make the move with us. By the time of our next move, he was married and starting a new life elsewhere.

The best regional conference we sponsored was held at South Padre Island, an upscale tourist destination on the Gulf coast near Corpus Christi, Texas. During this tour, Houston experienced a hurricane. It wasn't as bad as the ones today, but it did cause a good deal of damage and power outages. One of our brokers had a connection to a supermarket and I was able to secure a trunkful of bags of ice. I began a tour of our broker's offices in the Houston metro area, dropping off ice and donated sandwiches. We were not able to make a large impact on their losses, but I could tell they appreciated our efforts.

Being close to the Gulf of Mexico, the humidity was unnaturally high. I would shower in the morning, dress in shirt, tie and sport coat, and by the time I went to work, I already needed another shower. Another down side of the Houston area was, by the time we arrived the city was logging eight hours per day of peak traffic.

[107] John Caricature

Now I was deep in the heart of Texas and my western clothing fit right in. Our original office was in a different location, but the director had problems with the landlord. So...late one Saturday night, we packed up everything and moved to the location on the loop. We had great office space and a training facility. I was promoted to a newly created position, Director of Administration.

One of the brokers on the Houston's east side, ginned up a fancy do. He invited the Houston Oilers (NFL) cheerleaders to attend. About a half-dozen showed up and at one time I had a picture of two of the cheerleaders with me in the middle. Sorry to say that image disappeared years ago, so you'll have to settle on this picture of me done by a caricature artist that same day.

One day during the first winter, I was the only staff besides our secretaries who made it to the office. The weather was miserable with snow and low temperatures. The building heating system wasn't able to maintain a decent level. By mid-morning with the temp in the low 50s, I told the admin people to go home. No one could type more than a few minutes before the hands headed for the armpits. I called our boss at home and informed him of my decision.

I'm certain he didn't agree with me, but he said nothing. There were a lot of matters on which we disagreed. My wife, Pat and I were active with N.O.W. (National Organization for Women), and a letter to the editor supporting women's reproductive rights was published in the Sunday paper. On Monday, my Director told me he read Pat's letter; probably the only time in his life he read the editorial page. With little other comment, he described to me the reason why childbirth was so painful for a woman. To my surprise he let me in on the secret: it was all Eve's fault for bringing sin to the Garden, and that guilt was visited upon all females. He was a holier than thou–all hat and no cattle.

We had an underproducing office and the broker had a habit of asking embarrassing questions. My Director said he was going on a week's vacation, and he wanted that broker to be gone before he returned. Franchising is complicated business and federally regulated. I couldn't see any legal way to kick this office out of the system without cutting corners and without my fingerprints all over the mess. I thought it was unfair to that broker and my loyalty didn't extend to illegal activities. I did nothing and when the Director returned, he was not a happy camper. This was a time when I violated my own mantra which I cover in the next paragraph.

After a few more times taking the initiative and following my mantra that I'd rather be criticized for doing something wrong than for doing nothing at all, it came to a head. I became a statistic. My position was eliminated from their budget. I recall it being a Tuesday and I told him I would be happy to finish out the week. He said, today would be fine. I cleaned out my desk, said goodbye to more than one secretary who had tears in their eyes and headed for home.

I researched the local market for a new position while at the same time checked some Realty World contacts looking for jobs in other regions. The local regional manager from Century 21, invited me to check out his staff. They were, on a national basis, the largest of the real estate franchises, so I accepted the offer. Within a few days, I also was offered a position with the North Texas and Oklahoma Realty World region based in Dallas.

I told Century 21 I was moving to Dallas.

29) May 1983 – May1986, Dallas, Texas — Irving, Texas —Realty World, North Texas and Oklahoma

We found an apartment in Los Colinas located in Irving, Texas; just like the Irving Cowboys, the NFL football team. Yes, I know they went by Dallas Cowboys, but the stadium was in Irving.

I was again filling a dual role as Business Development Consultant and Trainer. I was also charged with supervising our BDC/Trainer who covered the state of Oklahoma. On my first visit to Oklahoma City, I audited the fellow in his role as trainer. He was well versed on the subject matter, but I had a problem with his teaching style.

The Realty World national trainer, the one who trained the trainers, told us presenting the course was only a small part of the job. We also had to be entertainers; if the students tune you out, they don't learn. He also stressed this point: 20% of your students will love you no matter what you do *to them*; 20% of your students will hate you no matter what you do *for them;* so, aim your efforts at the 60% in the middle. During those days, my knowledge was reinforced that people tend to learn around one of three senses: visual, aural and tactile. They often drop clues in their speech – I see what you're saying – I hear what you're saying – I can get my hands on that …

Part of being an *entertainer* was finding ways to present material in a manner that appealed to those three senses. Covering all bases was a challenge similar to tight rope walking.

Back to our Oklahoma City trainer. On the first day of class, he introduced himself to class wearing a shirt and tie and then climbed up on a high stool like the ones at a kitchen serving counter. Always praise in public and admonish in private. When the day ended, I sat down with him and described my approach in front of the class. I told him about dancing eight hours a day for five straight days while wearing cowboy boots; also mentioning Ginger Rogers.

I always began a class wearing shirt, tie, slacks and a Realty World sport coat. A few minutes after my introductory remarks, I would ask the classes' permission to remove my jacket, took it off and hung in on the cross arm of the replica of the Realty World five-foot white post and cross arm that was installed in front yards to hold the For Sale sign. Now, unburdened from the coat, I was ready to dance and entertain.

I politely told this fellow, that the first thing he had to do to become a first-rate trainer, was to get rid of the stool. On your feet, an instructor is free to move around the room and engage the students. He took my comments to heart and became a first-rate trainer. By entertaining, I don't mean a trainer has to be a stand-up comic, but having a few corny jokes, which generally evoke moans, break the ice. I usually had cues in my teaching notes to remind me to do just that.

This region was owned by an outfit in Minnesota, and they were looking for employee spots for *their own*. I didn't fit that definition in familiarity or temperament and once again I could see the writing on the wall. I was aware that a Savings and Loan company in Phoenix was looking at the Arizona region and the North Texas and Oklahoma region as a way of generating more real estate loans. They also had their eye on the then president of Realty World to head up this new venture. Again, it's good to be in the right spot at the right time. I'd met the man the S&L was looking to hire, so I corresponded with the Realty World national headquarters.

30) June 1985 – May 1986, 8787 E. Mountain View, Scottsdale, Arizona

My communications efforts paid off. The S&L was able to hire the president of Realty World to run their new venture and in turn he brought me on board as well. They completed the purchase of the two real estate regions of Realty World and I was named a vice-president of the sub corporation that managed this operation.

My boss was the entrepreneurial type and my attention to detail made a good fit. We were developing a computerized method for our brokers to communicate with the S&L to generate real estate loans. My boss asked me if he should learn the details of the computer and the software being developed. I said, no, that's my job and you need to concentrate on generating business.

Not that he couldn't operate a computer, but he didn't need to know how the guts worked, just to be able to operate it. At one point, the developer brought the package to us for a test run before going ahead. The fellow talked my boss through the process up to the point where he needed a password to continue. This was supposed to be a secure way for brokers to submit loan applications. The person demonstrating the program started to give my boss a password, but he said let me give it a try. My boss tapped several keys and nothing happened. Then he hit the ESC key and entered the secure area. The demonstrator said we weren't supposed to hit the Escape key. So much for secure. That lesson enforced my feeling about creating software: It's tough to make it foolproof, 'cause us fools are too inventive.

.I am reminded of a lady programmer presenting a seminar I attended. She said, "Never buy version one-point-oh (1.0) of any software package. It never works 100% right out of the box. Like the moon coming over the horizon when I was on alert duty that night we were nearly scrambled at Otis AFB.

31) May 1985 – April 1999 4102 E. Ray Road, #1164, Phoenix, Arizona 85044

We purchased a condominium on Ray Road, south of South Mountain in Ahwatukee, a suburb of Phoenix. There were about fifty units in our condo area. Good for knowing folks, not so good when shared expenses arose. Our unit was set back from Ray Road forty yards or so and the land formed a catch basin that could help in case of any flooding. Rain was scarce in Phoenix, but when it came it could be devastating.

The landscaping was beautiful and at the same time showed what was wrong. Residents loved Kentucky blue grass, olive trees and all manner of flora that wasn't native to Arizona. Desert landscaping was the way to go, but was ignored by too many. When we arrived, we could drive east on Ray Road, cross Interstate 10 into an agricultural area and the outside temperature would drop fifteen degrees. By the time we left, those fields were gone, replaced by concrete and houses and the temp change was a thing of the past.

There is a story in the Appendix about our cat, Callie, and the pretty catch basin in front of our home. It's a bitter sweet tale.

Back in the 1950s when I was in flying school, a high humidity was ten to twelve degrees. At this point in the timeline, humidity ranged to the thirties or higher. Man, do humans know how to screw with Mother Nature. We never seem to learn and only find out that she will win in the end and when it's nearly too late to do something about it.

REAL ESTATE FRANCHISES WANE

Franchises began to fade; I was mostly right when I said Century-21, Re/Max and Coldwell Banker would be the survivors. Realty World was the first international real estate franchise; it began in Canada and moved to U.S. before Century-21 opened its first Canadian office. I still believe that Realty World had the best training program for agents bar none. It became the victim of owners who thought it was a cash cow. It was not and it took patient cash to build and grow and in my opinion that was the last nail in the coffin.

At the same time, the Savings & Loan industry was facing challenges of its own. Looking back, I have to laugh—it wasn't funny at the time—at a group of fellow employees who were arguing and vying for a corner office on my floor. I mentioned that one should not covet an office when we might not have a job down the line. My prediction was correct.

My boss did his best to protect my position with the company. I got a different job but that did not last long. During that short period, I became an expert using Lotus 123, spreadsheets and was the go-to guy for those who had questions. That was a handy attribute since each "cost center" budget was in the form of a Lotus spreadsheet, and I was charged with managing over twenty "cost center" budget submissions. I had to be certain that each and all centers were included in the final roll-up for our department.

The answer was simple since each center had a four-digit designation in addition to its name. Since the numbers were summed as I added each center's dollar amounts into a single final product, that particular cell of information should equal the total of all the center designation numbers added together. I shared that approach with a friend who was in charge of the entire companies' budget process. Another feather in the cap of the Lotus Guru.

Again, the writing was on the wall. I was no longer employed. I should have been used to it by now, but I wasn't. It's tough being a job hunter in your mid-fifties. But…I do subscribe to the maxim that is does not make any difference how many times you are knocked down, as long as you get up one more time.

Shortly after I left the savings and loan job, a federal agency, I don't remember who had the power, but they changed the required reserves for S&Ls to meet. The decision affected the amount of cash and/or liquidity which was needed to continue safe business practices. Overnight, the percentage figure was raised to a level that not a single S&L met the requirement. I've often described that time as the day the savings and loan industry melted away like a lump of sugar in a cup of hot tea.

I found myself in a familiar if not a desired position.

A BRAND-NEW FRANCHISE

While perusing the want ads looking for a new position, I came across an ad that was worth checking. A brand-new real estate franchise was looking for people who wanted to get in on the ground floor. I didn't think they would be expecting to find someone with my background. I was right; a brief interview and I was hired to oversee broker relations.

I was writing my own job description as I went. If this outfit could get off the ground and succeed, I figured this was the bottom end of a rainbow arc. The owner kept the franchise contracts the member brokers signed in a safe in his office. Seemed like the thing to do at the time. Those contracts were the most valuable assets the company owned and should be protected from loss.

In a short time, I learned there was another reason. Franchising is a heavily regulated industry and the Securities and Exchange Commission (SEC) takes a dim view of any violations of their rules. One rule says you have to have a standard franchise contract and you must offer that contract to all comers; no changes allowed. For some reason, to carry out my duties I needed to review those franchise contracts. It only took the time to look at two of them to see a problem. I approached the owner regarding the differences in nearly every contract on file. He told me not to sweat it.

Along the way, I became friends with the fellow who handled the accounting for the franchise owner. Over lunch one day, he expressed his concerns about the owner's approach to creative accounting. He was concerned he could lose his CPA designation over it. An accountant without a CPA classification will be at the bottom of the salary scale. He left shortly thereafter.

The owner was a micro-manager and liked everyone within view. We had offices in two different buildings and there was an empty office in the second building near our agent training area. One afternoon I moved my belongings and furniture to my new office space. The boss wasn't real happy with the move and let me know. Damn, there's that writing on the wall again. Our agreement included a "probationary" period at the end of which salary would be discussed. Rather than "discussing" salary, he handed me the equivalent of a pink-slip.

Some business owners use money to make people, while others use people to make money. Unfortunately, this owner was one of the latter. I have no idea what became of him or the company.

Over my working life since the military, I've been caught in budget cuts, phased out, let go, fired, left behind, discharged, sacked, terminated, given notice … you name it, I've been there. I said to myself, I'm here again, time to get up again and become…

SELF EMPLOYED

[108] The Trash-80

I'll take a moment to provide my computer background since it led to this phase of my life story. Two years after Tandy Radio Shack released their computer, the TRS-80, lovingly called the Trash 80, we bought our first computer. It ran on TRSDOS, which stands for the Tandy Radio Shack Disk Operating System. We got the upgrade: a Model 1, Level 2, which sported 16k (16,000 bytes), of memory—twice the RAM in the Level 1 machine. To get anything in or out of the machine we used "special" audio cassettes. The "special" meant they cost more than the regular audio cassettes.

The computer was in the $600-700 range.

We spent most of our time playing Cranston Manor, a text-based adventure game. We typed in directions – go north, abbreviated as I recall—and the game responded with hints or situations, like "you're dead." Mark taught himself programming and had objects marching across the top of the screen and was able to shoot "death rays" at them.

Around the early to mid-80s, we acquired a Kaypro II (model 2), portable computer. It had a steel case and the keyboard unsnapped from the rest of the case and the unit offered a 8-9 inch monochrome monitor, in stunning green. There was no hard drive, but it did have two 5 ½" floppy disk slots. The operating system was CP/M which stood for Control Program for Microcomputers and had a suite of programs including WordStar a word processing program. This steel cased unit was a tank. I met a fellow at an airport lugging one of them behind him. He told me he'd just returned from an African trek where the 'tank' designation was truly earned.

[109] The Steel Battleship

A bit later in the 80s, I bought a Vendex, made in the Netherlands, and was IMB-XP compatible using an 8088 processor. It could run at 4.77MHz – and by pressing the "Turbo" button it would zip along at 8MHz – Wowie! It looked more like computers of the day; a separate monitor, a computer box and keyboard. It came with 20-40 mb hard drive and two 5 1/2" floppy drives. This baby came in around $1,000.

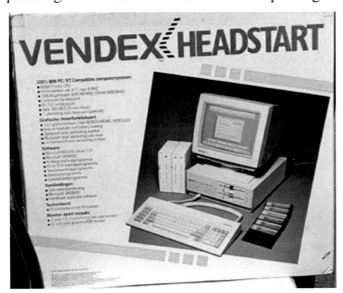

[110] My first IBM type computer

I checked the internet looking for this vintage computer, and the only decent picture I located was this one— an image of the carton it came in. Most pics showed the Headstart II which came with one 5 1/2" and one 3 1/2" floppy drives.

[111] Apple Macintosh SE

About the same time, we added an Apple MacIntosh SE. I can't find a price for this machine back in the day, but I think we went into hock for over a couple of grand. The internal hard drive was small, in the 20-30mb range; not large enough to hold the software packages I would be teaching. I added an external hard drive and paid a ton for 300mb's. I would delete one program and upload another from the external to prepare for the lessons I would be teaching. It also contained a 3.5-inch floppy drive, which is somewhat of a misnomer. The floppy disk had a solid plastic case, but the old name stuck.

The motherboard could accommodate 4mb's of RAM on four 1mb SIMM (single inline memory module) cards, but this one came with only 1mb in the form of 4 x 256kb cards. I purchased 4mb's of RAM and the tools to make it a DIY job. I read the instructions and picked up the case cracker tool. The Mac had a plastic case the tool would break the two halves apart. The notes said: you understand that opening the case voids any warranties. Oh, boy!

The mother board was lying flat at the very bottom of the computer. That meant I had to remove everything above it to locate the slots for the SIMM strips. Minor things like the cathode ray tube, yeah that thing that forms the monitor. Another warning: be careful not to break off the back end of the tube as you lift it out. This project was rapidly turning into brain surgery and I was sweating blood. Finally, I reached the bottom layer, remove the four old cards and inserted the new ones. Then I rebuilt the computer from the bottom up and cranked her up to test the RAM. It worked, and I took my first breath in the last half hour.

I asked someone what I could do with those four old SIMM cards. He suggested I glue then together and a stack of them would make a decent paper weight. Not much value in old memory cards.

NOW TO PUT THAT KNOWLEDGE TO USE

I located a lady in Scottsdale whose company had a contract with Motorola. She needed instructors who could teach computer skills to their employees I was teamed with a sharp lady who was a lead instructor. My job was to roam the classroom assisting students who were falling behind.

The Motorola plant was located in Mesa, Arizona and they produced computer chips for Apple computers. At the start of each class, we would ask the participants to introduce themselves by name and job description. Several said the worked in the flip-chip area. Besides knowing they worked in ultra clean rooms and wore gowns, gloves and masks. I couldn't stand it any longer so I asked what the flip-chip department did. They said the chips, and at this point they are rather large pieces of silicon, came down the assembly line lying flat. This area did just what the name said, they turned chips over so they could travel on down the road on their backs.

All of these classes used software for Apple products. As I recall, we taught word processing, spreadsheets and desktop publishing software. The Apple computers came with a Graphical User Interface (GUI), which Microsoft copied/pirated/borrowed (fill in the blank) and came out with Windows 1.0 in 1985. So, lot of my early work in word processing (Word and Word Perfect) and spreadsheets (Lotus 1-2-3 and Quattro) was in the DOS (Disk Operating System) environment programmed by Bill Gates.

At one point I was teaching three DOS versions of Lotus. They were very similar but there were enough version differences you couldn't fake it. I needed all three to learn them. When they came out with the latest version – this may have been the Windows version – I called the Lotus company; I told them about the number of Lotus versions I owned in order to teach and asked if they had a teacher's discount. The answer was NO. I told them that was tacky, bordering on plumb shabby. They basically said was that's tough.

Word Perfect had a similar attitude; i.e., our s—t don't stink. That's no doubt why today you know Word and Excel and if those other two are even still in business they are decades behind.

COMPUTERS IN A DIFFERENT VENUE

During this period, I put a resume on file with Mesa Community College, aptly located in Mesa, Arizona which is a suburb on the east side of Phoenix. MCC had a student population of around 3,000 and was second in size only to Miami-Dade. It was nearly a year since I made the application, so I called the business department to see about updating the paperwork.

The business office told me the head of the department would like to speak with me. Within a week, I was sitting in his office expecting the usual employment type interview. The appointment didn't last long, and I walked out of the office with a stack of books under my arm and a class assignment that would start in a couple of weeks. I would be teaching a computer 101 course that covered a wide range of subject matter but not a great depth. I learned a lot of computer history myself. Most of those in my classes were freshman with a sprinkling of older students; plus, an occasional individual who knew more about computers than I did and just needed the extra hours for graduation.

I believe the goal or objective of education is to teach students to think, and if they gain some knowledge along the way, even better. There was the usual before-something-new anxiety. I'd done a lot of training, but never in such a formal setting. Since I didn't have any education background, I was required to attend a class to qualify to be an adjunct faculty member. Adjunct staff is a euphemism for; we take up the slack so the full-time faculty can make the big bucks.

That class was scheduled for five or six weeks on Saturday mornings. The facilitator opened the first session with: this will be the most boring time you'll ever spend, but it's required, so let's get to it. He was right. I completed the course and was now a qualified instructor for the state of Arizona.

I prepared a syllabus, read the text book and practiced each class before heading to the campus. I used our "computer" room as my classroom and the printer as my podium. The printer was an Okidata located between my PC and Mac computers. Remember dot-matrix printers? No toner or ink cartridges, just a typewriter kind of ribbon. It could create near-letter perfect copy by printing a line of text, do carriage return without shifting down a line, and then offsetting a pixel or two and reprint the same line. It was decent for the day, but stone-age compared to today's equipment.

I stood in that computer room and for the first semester, I presented each and every class to the bookcase against the wall. I felt confident at my first for-real presentation approached and enjoyed all the time I taught at MCC.

I developed a grading system which would allow top students to achieve an "A" without taking the final exam. Build up enough points for an A before the time for the final. Others could earn a top grade if they had the point count, but would need a reasonable grade on all exams including the final. Attendance was worth a point a day, so it paid to show up. After the first couple of weeks, I could associate faces with names and told my class, that from now on I would take a silent roll call. Then I would announce that "roll call is complete," and if they arrived late and didn't hear the announcement, they would have to report to me after class in order to earn the point.

I also had a way to handle the perpetually late student. As they walked into the classroom, I turned my attention the opposite direction and announce, "… and that's the most important thing you will learn in the class all semester." When the recalcitrant would turn to a neighbor and ask what I just said, most of the students would go with the flow and not tell them anything.

CHALLENGES BEYOND TEACHING

I began one semester seeing only twelve names on the class roster—about 35% of the normal class size. My mind was reeling running over possibilities to one-on-one time and projects a student might want. I was just about to announce this to my students, when the dean of the business department came into my classroom, leading a herd of folks. He apologized for the mix-up and handed another fifteen or so off on me. It was a fun class but not the one I imagined. Roll with the changes. Years in the military taught me that. And we proved our own adage: the crews will get the job done, despite the staff.

In the middle of another semester, I met my most serious challenge. With about five minutes until the class was to start, one of my older students approached me. From previous discussions I knew he was a Vietnam vet and was married with a child. This day he confided to me that his wife had just left him, that he was now a single parent and…he felt like taking his own life.

I looked at the class to tell them that I would be back shortly and to review the day's lesson. I took this young man, younger than me, to the business department office. He sat in the waiting area as I kept an eye on him and asked the department's secretary to join us. I gave her a quick and dirty background on what had just happened including the reference to suicide. From past experience, I knew this lady was sharp, knowledgeable, capable and would immediately grasp the urgency of the situation. My instructions included that she should keep a close eye on my student, get in touch with the school's counselor and not let him out of her sight until the counselor took charge of him. I later learned she did her job and the therapist took this man in tow. I never heard what happen to my student, he didn't return to my class. My fervent hope was that he realized that suicide is a permanent solution to a temporary problem and led a fantastic life with his child.

The secretary and I understood how serious the situation was. I'm not sure how close my student was to suicide, but I figured his comment was a plea for help. He was overwhelmed and had no idea of where to turn. No matter how close to the edge he was, a comment about suicide should never be ignored or passed off as superficial.

WHAT COMES NEXT

When we moved to the Phoenix area, where our first child was born, we thought this might be our last move. Unfortunately, utopia changed during our fourteen years here. We lived south of South Mountain, and in the early days, we drove from our condo on Ray Road, east, cross I-10 and surrounded by fields of crops. The temperature would drop ten degrees as we passed I-10. Back further while in flying school, we would drive from Phoenix northwest to Glendale and smell the onion fields along the way.

By the late 1990s, those onion fields were gone; the fields east of I-10 were gone; people trying to emulate their home states planted olive trees and Kentucky blue grass and destroyed the desert environment. Back in the day, a high humidity day was twelve percent—now it was running forty percent or more. Pat's office was on the sixth floor of a building a half-dozen miles from the center of Phoenix, and the elevation there was a few hundred feet above the center of town. The bowl that formed Phoenix was a heat sink. The few and far between rain storms split and went around the city, not over it.

With nothing but bone-dry land, Habu's were numerous, especially between Tucson and Phoenix. A Habu was a huge, great big dust storm; often a hundred feet tall and miles wide. They were visible for miles, usually accompanied by TV station weather warnings and which obscured visibility to zero.

Back to Pat's office on the sixth floor. On a bad day, she could look out her window and see a brown haze like a lid on the bowl that was Phoenix. That got us thinking about yet another move. Pat and a girlfriend went on a trip to Arkansas. That friend always wanted to visit War Eagle, a huge, crafts show in northwest Arkansas that lasted for days. Pat returned with glowing reports of the area and said we needed to take a look.

A short time later, we drove to Arkansas, did some sight-seeing like Hanging Judge Parker's six-man gallows in Fort Smith and down through Hot Springs and Little Rock.

Just north of I-40 near Fayetteville, Arkansas, we stopped at the Pea Ridge National Military Park, memorializing a Civil War battle in 1862. As I looked out over a scattering of cannons and the vast expanse of green where the fight took place, the first lines of a poem formed in my mind: I hear the crack of musketry, an' I hear the cannon's roar, I never figured that fightin' was, such an awful chore;…

I titled that poem, simply—Pea Ridge. I don't write much poetry, and I prefer rhyming poems like those of Robert Service, the Bard of the Yukon (you may remember: The Shooting of Dan McGrew and The Cremation of Sam McGee). I'm proud of this effort; I touched something deep within me—still not exactly sure of what it was—and it took me four years to complete my poem. I will include it and a "back cover blurb" about it in the Appendix of this book.

Traveling up state Highway 7 from Hot Springs, we came upon Hot Springs Village. We almost passed it by; but stopped and liked what we saw. HSV is a gated community with a population around 12,000 full of curvy roads and hilly piney woods.

Back in Arizona, we discussed a move and set a goal. We would move to Arkansas within one year. We met that goal; almost a year to the day we sold the condo, moved to Hot Springs Village, rented a three-bedroom, supervised the construction of a house, and closed the sale on our new home.

RETIRED, RETIRED

I gave notice to Mesa Community College and quit my last day job as adjunct faculty. I'd lost track of how many times I moved from one career path to another. But this was the final and very last time. I was truly retired…retired.

Part of our trip prep was to acclimate our cat Lexie—that was short for Lexus, because she had a sleek coat she was named after the luxury car—to a car ride. We took short trips often stopping at a McDonalds for ice cream cones and sharing ice cream with our feline. We learned a collar was useless for Lexie; she slipped her head out of it. A snug fitting harness did the trick.

With our household goods loaded into a moving van, Pat and I headed out in two cars. Again, this was the time BCP—Before Cell Phones. I purchased a pair of walkie-talkie units so we could stay in touch car-to-car. We headed north on the interstate to Flagstaff, hung a right in a light snow shower and pointed our noses east. We planned for a three-day trip, figured that was the limit per day for the three travelers. First night, motel, second night with our daughter in Oklahoma City and the third day to Hot Springs Village where we would become Arkies.

Later I wrote a short story about Lexie and the trip, titled The Grand Adventure – First Person Feline. We got a lot of things right, but the story relates those aspects we got wrong. Among them, was the disposable litter boxes we purchased for the trip. Poor Lexie thought they were a great spot to nap, but that left her with no place to deposit her waste material. If it had occurred to me, I would have saved a Zip-Loc full of residue from her old litter box. Lessons learned, too late.

I entered this story in a contest at a writing conference and it took the Grand Masters Award for the entire conference and put a hundred bucks in my pocket. Buoyed by that experience I looked for cat magazines that might be interested in it. The response was underwhelming, but one editor did reply; she said, "We do not publish articles written by cats." I think she missed the point of the story as a learning exercise. That's a typical ride on the rollercoaster of getting something published.

The last leg of the trip, Oklahoma City to Hot Springs Village was the worst. All the way across I-40 it poured rain so hard it was difficult seeing the hood ornament. Even though the east and west lane were separated, they were close enough we caught spray from oncoming semi-trucks that reduced visibility to zero. I was leading and would warn Pat when a semi was approaching.

At Russellville, Arkansas, we left the interstate and turned south on state Highway 5 to Hot Springs Village. The weather cleared and the rain stopped as soon as we made that turn; a good omen for the future we hoped.

32) May 1999 to April 2016, Hot Springs Village Arkansas 71909

3 Alarcon Way, Hot Springs Village, AR 71909 (rental) May – November 1999

This was our base of operations as we shopped and contracted to have a house built from the ground up. We never imagined how many decisions would need to be made. We picked a plan that provided as much living area as possible without wasting space on things like hallways. Decisions included picking the color of the aluminum siding, whether there would be windows in the garage door and landscaping. I never enjoyed shoving a lawn mower around, so we opted for rocks instead. The front yard included a huge boulder; at least three by five feet and two feet tall.

[112] 42 Vega Lane

The construction boss said we would have paid several hundred dollars for that rock. Glad nature left it there for us.

As I mentioned before, our closing date was virtually a year to the day we said we would be ensconced in Arkansas. That date was almost missed, we had to hope the construction folks would show up to work rather than going hunting. Hunting in Arkansas is a serious situation and carries a high priority

32) 42 Vega Lane, Hot Springs Village, AR 71909, December 1999 to April 2016

BEGINNING A WRITING CAREER

I dabbled as a writer from my teens, mostly westerns and space travel, to the present. Way back in Bellevue, Nebraska, I wrote a scene, around 10,000 words, about a tanker pilot who saves a fighter plane in Southeast Asia. I thought it was pretty good, and incorporated it into a novel as a flashback to set the back story of the character. The members of the critique group I met with here in Hot Springs, Arkansas, suffered through it and then leveled with me. Waaaaaay too much flashback. They were right, and I ended up cutting about 9,800 words out of it. That's how a writer learns. Sometimes you have to kill your children.

My first introduction to critique groups was in Phoenix. There were five of us and I learned more about writing from these folks than I did from the few writing classes I attended. A couple from this group self-published their books. I was still struggling with my first mystery. When I left, the group presented me with the wooden "book' shown here. The plaque reads "A writer is someone who writes every day. Best Wishes." That sums up what a writer does, and it's the reason I borrow this line on notes to author friends. It's by Natalie Goldberg writing teacher and author of "Writing Down the Bones" – Keep the Pencil Moving.

I abandoned those earlier genres and developed a female protagonist, Casey Fremont, who gets into the middle of murder and mayhem. In Arizona, I attended a day-long seminar with Val McDermid, a British mystery writer. She told us it took her two years to develop a story line for each of her books. I scoffed, to myself thinking I could do it in less time. She was right.

[113] Gift from AZ critique group

I began developing Casey Fremont in Arizona and finished nearly two years later in Arkansas. It took that amount of time to develop the character, and now I faced plots to develop. That's the back story of why Casey lives in Little Rock and not Phoenix.

I located that critique group which met in Hot Springs every Monday. All six of us were serious writers of long fiction, i.e., novels. It helps when members are working on the same type of writing even though the subject matter may cover a wide range of storylines.

The Village also had a group called The Village Writers Club. I did a number of presentations to this group as well as for our local recreational library. I developed a fast friendship with quite a few from the VWC, and I am still in touch with several of them. One of my toughest chores when we decided to move again was to say goodbye to this group. They reciprocated by presenting me with the wooden plaque shown here.

[114] Gift from HSV Writers Club

LIFE BEYOND BEING AN AUTHOR

Years of being on the voter rolls finally caught up with me, and I received a notice to report for jury duty. That required me to report in each day to see if I was needed. After a few weeks my name floated to the top and I showed up in the county courthouse around thirty miles to the east in Benton, Arkansas. The case involved a lady in her sixties, who was a heavy smoker even though she was diagnosed with COPD. She was hospitalized and unable to eat; the staff put a feeding tube down her throat; missed the stomach and into the lungs. She drowned from the nutrients she received.

Her doctor was on trial for whatever, I don't' remember the exact charge, for not checking the tube insertion and for not supervising her nurse who did the tube insertion. During voir dire, the doctor's lawyer began talking about the victim, that she was old, she probably didn't have much time left because she had COPD and was still smoking. I supposed he was trying to slip in mitigating circumstances even before the trial started. After that he began asking me questions. I answered and added that I could not believe he broached the lady's health. I said something like this: I don't care whether the woman had fifteen years or fifteen minutes to live, I can't believe you even brought up that line of the story. He began to ask me if I could set aside those facts and render a judgement, when the judge interrupted and said, "You are excused." I gathered up my belongings and left the courthouse. Thus ended my stint as a juryman.

We met with a group of folks calling themselves Unitarians. They were friendly and had a view of life and religion that was similar to ours. For quite a while we met each Sunday morning in a meeting room at one of the Village clubhouses. I've always had the feeling that every religion I'd come in contact with, was a shared desired they wanted for control imposed on members. Perhaps the best way to describe Unitarian Universalists was a cartoon this group displayed in the meeting room. The background displayed a typical church, large and with a towering steeple. The front lawn sloping down to the sidewalk and a wooden sign staked into the grass. The sign read: BEWARE OF THE DOGMA.

Steeples are another matter too. Seems each church does it's best to outdo its neighbors with the largest steeple possible. Since most churches are overseen by old white men, I think the steeple became a phallic symbol—My steeple's bigger than your steeple. I hope I haven't offended any old guys with large phallic symbols.

The time arrived for this group to look for a permanent facility. That was a long contentious process. We got it done and moved into a church building the congregation of which build a new, bigger building, with a proportionately sized steeple. One of our members, a retired minister from another denomination said, "If you think that was tough, just wait until you try to hire a preacher." He was right, but by that time Pat and I were on our way out of town.

I loved that group. They were: a retired minister who I just mentioned; people looking for something and not knowing what; lapsed Catholics; atheists, deists; skeptics; spiritualists; and some just looking for a group to socialize with. The point of Unitarianism is that all those wide and diverse folks were welcome.

If you are religious and happy with your situation, you may want to skip this paragraph. This is a summary of a talk I presented to the Unitarians I met with. As a child I was taught a prayer to say each night. I dutifully did it with little understanding and mostly by rote; Now I lay me down to sleep, I pray the lord my soul to keep; if I should die before I wake; ... one night it hit me —Whoa, Hoss. You mean I might not make it through the night. Man, that's bat crap crazy. Then the church says, now that I've scared the Bejesus out of you, here's what you have to do to escape the fiery future—just follow this list of rules, a little less than a mile long, and after, AFTER you die you have a home somewhere, doing something...

I'm doing my best to be what I refer to as being a spiritualist, to live the best life I'm capable of and leave this planet a little better off than when I arrived here.

NINE-ELEVEN

[115] Canada Vacation

Rather than a treatise on that terrible attack, this part of my story is how that awful event got in front of a trip we had planned. Nine-eleven occurred on a Tuesday; our Canadian Rockies train and bus tour was scheduled to depart Eppley Airfield in Little Rock, Arkansas on September 12th, a Wednesday. We all watched the horrific images of planes plunging into buildings, all the time wondering what effect it would have on the airline industry.

Our travel agents were unsure as well, but told us they planned that our departure to Vancouver, British Columbia would be a go. We dutifully boarded a bus in Hot Springs Village around Oh-Dark-Early and headed up the interstate toward the airport. During that ride, confusing phone calls were received by our travel escort. On as scheduled, Yes…No. No flights in or out of the airport, Yes…No. When the bus pulled up to the departure terminal the definitive answer was…ain't no planes flying today, folks.

The bus turned around and began the leg back to home. We convinced the driver to make a schedule deviation and hit a McDonalds fast food emporium. They were not expecting a bus load of diners at that hour of the morning, but did yeoman's duty and got the job done.

Back at home we awaited developments wondering if travel insurance covered terrorist attacks. The country reopened the skies in a couple of days and we reboarded that same bus on the following Friday or Saturday—my mind's a little foggy on the day.

We picked up the trip only missing the first couple of days and enjoyed the view of fantastic landscape, upscale hotels at Lake Louise and Banff and meals on the train so often I lost count. After several days we switched from the train to a bus to continue. They warned us, but we didn't listen; takes a while to get your "sea legs" for the transfer from a constantly moving vehicle to walking on solid ground.

Pat and I boarding the Rocky Mountaineer.

[115a] Canadian train

One memorable moment came when the bus driver, who promised to show us wild life on this leg, finally eased to a stop half off the roadway. He told us to check the left side windows and look down at the asphalt and added: see that brown pile of material sitting in the middle of the road? A bear has been here not long ago. We did see some actual wild life but the bear droppings were a highlight.

In Calgary, we did a bit of sightseeing and the most impressive sight was a Royal Canadian Mounted Policeman, a good-looking young man in his dress uniform doing PR work with the locals and the visitors. I say good-looking, because most of the ladies on the trip seemed to think so.

We returned home and I had a small amount of Canadian currency left. I stopped by my bank and they declined to convert it to U.S. dollars because they had a minimum amount and I was way below the number. As I recall, I had a few Canadian Dollars (CAD) and some coins; Loonies and Toonies. The Loonie is a dollar and nicknamed for the loon on the reverse of the coin. Toonies are two-dollar coins.

So, what to do with a few dollars no one would take as legal tender? I thought of an online friend; I knew him from multi-player flight simulator flying—we both flew with DC-3 Airways, a virtual airline dedicated to the old Gooney Bird. Ron lived in Saskatchewan, so I asked him for his snail-mail address, and I bundled up my Canadian fortune and sent it to him. Ron let me know he visited his meat market, bought a large soup bone and presented it to his dog, Sasha. We shared laughs later on as he described his pet pooch burying the Achor-Bone in the back yard and then going out the next day in search of the missing snack.

I met a great bunch of guys on DC-3 Airways, "When Flying Was Fun." I had friends all over, beyond the U.S. (Bob and Mark). I knew and flew with fellows from Germany (Hartwig), Britain (Norman), Canada (Ron), Australia (Ray) and many more. Till we meet again on the "Silver Wings" page.

FIFTY YEARS AND COUNTING

For our 50th wedding anniversary, we decided to travel to Oklahoma to visit our daughter and her husband.

Image 118 shows us in the backyard in Edmond, Oklahoma.

[118] Author & Pat 50th wedding anniversary

[119] 50th anniversary meal with daughter (Kathy) & husband

Here we are a few hours later enjoying a good meal and wine at a local eatery.

Lastly, a poem I created for this date.

50th Anniversary – June 18, 2005

To Pat — An understanding and loving partner on our Golden Anniversary

We have traveled a road, a half century long,
Past the quiet verses of a peaceful song,
Through quiet times then past the rants and the raves,
Over tranquil ponds and beyond the storm filled waves.

An hourglass shape is rearranged and has shifted,
My youthful waist has grown and expanded,
Silver threads find their way into your strands of gold,
My scalp needs protection now, from sun and from cold.

From coast to coast, traversing states of land and mind,
We traveled the highs and lows, we found the ties that bind,
Children three, they grew, they left for lives of their own,
Times spent together and times apart, all of these we have known.

ENSHRINED IN THE LIBRARY OF CONGRESS

A Village friend of mine, Jeff Meek, put me there. As a volunteer, and with a lot of work, Jeff videotaped vets telling their life stories and submitted the finished product to the Library of Congress – Veterans History Project. He recorded me in the early 2000s and he already had over 200 video tapes to his credit. Thank you, friend. You can find this effort at:
> https://www.loc.gov/item/afc2001001.95861/

AND BACK TO WHAT I WROTE

I've already mentioned, "The Golden Days of Radio – and Box Tops." It was published in "Good Old Days" magazine and was my first paid gig. A whopping $40… and I count is as the beginning of my professional author days.

[120] Autor book covers FB banner

Check the Appendices for this one.

My three Casey Fremont mysteries were published by a small press in north Arkansas. We later parted company, on reasonably friendly terms, and I republished those mystery novels under my own imprint—Acacia Imprints.

A small town south of Hot Springs Village put on an outdoor event which was fun. I made it to Searcy, Arkansas for quite a number of their day long seminars. A great batch of people and great writers. OWL, Ozark Writers League, holds quarterly conferences in Branson, Missouri, and I met another new batch of writer friends.

I have 50-60 plus short stories in my portfolio. I've included several in the Appendix to this book.

The AWC (Arkansas Writers Conference) is an annual event held in Little Rock. I lost track of the number of these events I attended. Many familiar faces and a few new ones each year. The Hemingway–Pfeiffer Museum and Learning Center (HPMLC), located in Piggott, Arkansas–about as far up in the northeast corner of Arkansas and still be in Arkansas, holds an annual writers' retreat for adult writers. Yes, that's Ernest Hemingway you know. Pauline Pfeiffer was his second, but not the last, wife of Hemingway. Thanks to her wealthy uncle, Ernie was able to become an expatriate in Europe. He wrote, at least part of, "A Farewell to Arms" in the famous barn studio on the site. It was inspiring to be there, looking at his typewriter and other trappings around us.

[121] Author at Hemmingway retreat

The staff and facilitators at HPMLC, are a fantastic group and I count them as friends forever. They've also expanded to include a number of activities for young writers as well. The size of the group was limited to a eight to ten, and it was a delightful and interesting week of writing and learning exercises. I will always remember our facilitator, Rob Lamm, explaining what onomatopoeia was and how to use it. I consider Rob one my most important writing mentors...thank you, my friend.

OTHER DIVERSIONS

A WW II Boeing B-17, Flying Fortress owned by the Collins Foundation came to Hot Springs, AR. They offered rides, but I settled for a tour of the plane.

The original plane was repainted to replicate the 91st Bomb Group's **Nine-O-Nine**. That plane completed 140 combat missions.

[122] Boeing B-17

[123] John in Cockpit

The author clambered up and into the cockpit of the bird.

Unfortunately, this aircraft crashed in 2019 and was destroyed.

A bit of Deering-do…while still in on-base housing, I purchased this Honda 100. Actually, it was a 90-cc engine and they puffed the advertising.

It was built like a dirt bike, but offered enough equipment that it was street legal.

Old adage: there's them that's laid 'em down and them that's goin' to lay it down. I was still upright after a year and a half, so I sold my little red Honda.

[124] Author and motorcycle

PILOT CLASS REUNIONS

USAF pilot training class 57-I had two and a half reunions—2004, 2007 and one that was cancelled

I was webmaster for a site dedicated to those first two. Figure 125 is the actual cockpit of a KC-97 attached to the Solo Restaurant in Colorado Springs, Colorado in 2004

[125] Author and wife having lunch

Family illness prevented us attending the 2007 reunion and as I said the third one didn't get off the ground.

MURDER IN SPA CITY

Hot Springs, Arkansas was nicknamed Spa City along with some other nefarious activities. I was a co-facilitator and emcee for a conference we called: MISC – Murder in Spa City. We were betting on the come hoping we could offer enough good material we could attract the numbers that would put us in the black. It was tight but we did it.

I began scrounging for big name speakers who would bring the people to Hot Springs. Velda Brotherton, a successful Arkansas western writer said she would be happy to be there. I knew she would draw many from northwest Arkansas and Branson Missouri. I also knew Laura Parker Castoro, a prolific romance writer and was just starting her mystery books. Her fans would fill seats as well.

I looked around for a keynote speaker and presenter. Back in those days, Charlaine Harris was moving onto the national scene and was living in Arkansas in those days. Her mysteries and Southern Vampire Mysteries, featuring Sookie Stackhouse had a huge following. Her vampire books were adapted for television and were due for release. About that time, the TV writers' strike hit and Charlaine's television hits were delayed and she graciously accepted the amount of money, definitely not a princely sum, we had in the coffers for that slot.

[126] (L-R) Castoro, Brotherton & Harris

Featured speakers and the Keynote speaker (Harris) at Murder in Spa City conference

As Emcee for Murder in Spa City, I was busy keeping things on track. Thanks to many members of the Writers' Club of Hot Springs Village for making my job easy.

The date/time stamp reads: 04/04/2000

[127] Author at mike as Emcee for MISC

33) April 2016 to present - 19301 Seward Plaza, Apt. 218 (later 106), Elkhorn, Nebraska, 68022

Sixteen years in Arkansas was the longest stint at the same address we managed in the sixty-plus years of our marriage. It was time to pull up stakes and move once again. We picked Nebraska where we would be nearer to part of our family.

In October of 2015, we flew to our area of choice spending a week in Omaha checking out possible places to live. Due to Pat's outstanding planning, we made the rounds, examining retirement communities. We each picked a number one and a number two and compared notes. Our picks matched and we contacted Elk Ridge Village marketing and signed a contract holding a specific apartment. Now to get it all done and make the move by April or May, 2016.

Again, planning paid off; our home sold quickly and we piled into an overloaded automobile and headed north. We arrived on April 30th.

ADDING MORE WRITING TO THE CATELOG

I added a pair of thrillers, featuring Alex Hilliard, a USAF pilot, to the Acacia Imprints' catalog. Next came "The Iron Pumpkin," which is a memoir about Rivet Ball, an RC-135S, PHOTOINT, a one-of-a-kind recon bird and the fateful night I took her over the cliff at Shemya AS, Alaska.

I was working on this autobiography back in Arkansas. If I ever get to the end, I'll add it to my catalog. All these mysteries, thrillers and the memoir are available through Amazon, and details can be viewed on my website: www.johnachor.com

Computer flight simulators have been an interest of mine since the late 1980s. in several cases, I combined writing and flight simulators. "Follow the Bouncing Ball," a story included in the Appendix is a good example. I took a sabbatical from flight sims around 2010 and then returned in 2018. I got involved in Digital Combat Simulator (DCS), which has a steep learning curve for the software itself and the planes themselves. I've concentrated on the F/A-18 Hornet and the F-16 Viper planes. The Hornet is carrier capable and I'm doing my best to catch the Three-wire. Mostly, I'm crashing or if I'm lucky, accomplishing Bolters.

That background brought me to a new endeavor, creating videos about DCS and posting them on YouTube. I invested in a moderately priced video editing software program. Now I don't have to do fifteen to twenty minutes correctly the first time out. I can edit, rearrange clips and add voice-overs. At this time, I've posted four or five vids to YouTube. Along with those, I decided to do a series about the planes I flew during my twenty years in the USAF.

I located all the airplanes, mostly free-ware, designed short flight plans for each, and recorded the video which included me narrating the flight. I burned the videos to a DVD, designed a label and mailed them to our children.

In each video, I recount stories that occurred during the time I was flying each airplane. It was a fun learning project and to date, only my family has had them foisted upon them. ☺

FOLKS I'VE NEVER MET IN PERSON

Shortly after arriving in Nebraska a local reporter contacted me. I don't remember how we came to know one another, but Steve Liewer is the military affairs reporter for The Omaha World-Herald. He was aware of the accident involving Rivet Ball and wanted to do an article about me for the upcoming Veteran's Day supplement. I got a full page spread in the special supplement in 2016.

Steve introduced me to another military pilot and author via email. Robert S. Hopkins III, authored, "The Boeing KC-135 Stratotanker: More Than a Tanker." In this tome, Bob chronicles every 135 airframe that came off the assembly line, how it lived its life and how and where it died if appropriate—of course, many of the planes are still in the air. He also covers all the reconnaissance versions and modifications to the bird. Bob flew over fifteen variants of the 135 family. I flew two of the planes he covers and got several mentions in his book. I always love to see my name in print, thanks Bob.

I ran into another Robert on line. Robert Archer, lives in Great Briton and along with other talents, belongs to a group of plane watchers who specialize in 135 recce birds. When we met online, he was in the final stages of completing a book called "Super Snoopers." He was aware that I flew recon missions from Alaska over the Artic Ocean landing at RAF Upper Heyford and returning a few days later. He asked if I could let him know what it was like flying in those days. I sent him a pile of prose and he edited my words down to a couple of entries he included in his book. Thank you, Bob.

If you search Amazon about books that might fit this genre, you're likely to find one called, "Hognose Silent Warrior," by George F. Schreader. He was a linguist flying with the back enders who did most of the work on recon planes. I took an RC-135D from Alaska to Kadena, Okinawa where we flew missions into the Gulf of Tonkin using local language specialists from the USAF Security Service. I figured we might have flown together so I got in touch with him. His publisher forwarded by request to him and he got back to me by email. In his book, he provides a ton of background on how and why the Vietnam war happened. George also provides a history and story of how a foreign language expert is trained and what life is like flying around Southeast Asia in the aft part of the bird with very limited access to a window. I am enjoying his book and learning how they lived back there while I was upfront driving them around in big circles.

MIGHT AS WELL TOSS IN THIS LAST BUMMER

When COVID 19 hit the United States, despite a total lack of a coherent plan from the very top, I took every effort I was aware of to avoid the bug. Our retirement community went into a virtual lockdown. Lunch, the main meal of the day was delivered to our rooms. Kudos to the kitchen staff for completing their daily routine covering three floors in two buildings— on wheels

I nearly made it, but in December of 2019, about six weeks before vaccines became available, the EMTs hauled me off to the hospital. I remember muttering, "I don't want to go," as a nurse, my wife and the ambulance crew said: Oh, yes you are.

That was on the tenth of December and for the next two weeks, I remember nothing. Not a hint of where I was or what I was doing or what was being done for me. Bits and pieces of the next period bounce around in my head. I refer to that phase as a deep, dark pit of living hell.

Mostly I remember taking a swallow test and fearing I wouldn't pass and have to remain in the hospital. I passed and on January 4th I was transferred to a specialty hospital—my guess was that it was among a batch of pop-ups that took advantage of a lack of care facilities. I've told people if they want to send you there, run away as fast as you can. This was the worse hospitalization I've ever experienced. I could add another double novella about them, but I won't. The worst three weeks of my life, bar none. On January 25th, I moved to a rehab center. My thanks go to the wheelchair van driver who made it through near blizzard conditions to my new digs. It wasn't a fun place—way better than the hole I just left, but the physical therapists did what they are good at. I remember feeling the elation when I took 200 steps down the hallway without having to stop and rest. A trip around the world starts but with a single step. I was happy.

Being the gentleman I am, I shared that lousy bug with my wife, Pat. Fortunately, she didn't have the worst of it, but related pneumonia put her in the hospital and rehab. A bit of good luck, she and I were sent to the same rehabilitation facility. We ended up on the same wing, same floor so we often shared meals together in the small dining room.

I set a goal to be out of rehab, my birthday, February 27th was the date. I beat it by ten days and headed back home. Pat and I were both released on the same day. There was some trepidation facing life on our own. After nine weeks—two months plus a week—of structured time and activities for me, we would be responsible for handling our own affairs. I think we did and are handling that aspect well.

Side effects: loss of hair and memory problems are among them. Mostly, they are short term, but long-haul problems do exist. And it's hard to tell if fatigue and lack of energy fall in that long-haul category or are due to adding another year to our life span. We continue to plug away and accept each day as it comes.

WHEN DID MILITARY LIFE BEGIN?

Earlier, I indicated my career started in high school. In retrospect, my military life began in 1941. My dad was a member of the National Guard's 38th Infantry Division; the Indiana Cyclone Division. He was a master sergeant, but when they were called to active duty his position, in the expanded table of personnel, called for a captain. He received a direct commission to the rank and got to skip the brown bar/silver bar steps. After Mississippi, he was transferred to the 99th Infantry Division, the Checkerboard Division whose home was Pittsburg, Pennsylvania. They were shipped to Europe in 1944, first to England then onto mainland Europe after D-Day.

I still have a small (4" x 6" which folded in two) "leatherette" picture folder which he carried through his time in the war. Each image was protected by a clear plastic sheet and on the back, they read: Me–10 years old, March, 1944. Mom 39 years old, March, 1944.

He didn't talk about the war to any great extent, but I will share a few of the stories I know. During peace time, the national guard was required to recruit a certain percentage of Category 4 people. In today's jargon, they would probably be referred to as Special Ed folks. When the time arrived to go overseas, those Cat 4's were to be released from active duty. The headquarters where dad served had a young man who fell in that group. When told he would be sent home, he was devastated. The staff looked around and found their own loop hole. Even back then, the troops would get the job done in spite of the staff. The young man went to Europe along with his division, fared well in the environment and survived the war.

Dad said they were briefed to be on the lookout for enemy fighters. They were told if you saw the machine guns winking orange at you, the weapons were aimed at you. In this situation, they were far enough behind the battle line, they were wearing the Class A uniform, Pinks and Greens. The blouse was a deep olive green and the slacks had a definite pink hue, hence the nickname. Walking to work one morning, he heard the sound of a fighter plane, looked up and identified it as German. His only cover was the gutter, which offered little protection. From his prone position, he could see the guns on both wings winking orange. The plane passed and he got up thinking he was lucky the German pilot was a lousy shot. When he arrived at his office someone said his pants looked to be on fire. Dad looked down at his leg, a saw that a bullet had passed between his legs and singed his trousers.

One of his additional duties was to care for the Division General's trailer. Many of the top wheels used a mobile unit as a headquarters on wheels, and in the field provided sleeping quarters as well. They were in a small town and the general was using other quarters. Dad had the trailer backed into a "V" in a masonry wall surrounding their compound. That night, a single German airplane flew over; dropped a single bomb and guess where it landed. Yep, you beat me to it. That bomb landed square on the generals' trailer. Good planning, but no good deed goes unpunished.

The Checkerboard Division was in the thick of the Battle of the Bulge. Fearing their headquarters would be overrun, they were ordered to do a bug-out. They were in charge of a large amount of currency (probably "script" to be used to replace local money). There wasn't room on their vehicles to haul it away. They left a single officer sitting on top of the money which was laced with thermite bombs with orders to burn the money before the Germans could get to it. When the battle was over, the German advance didn't get to the area of the money and they found that single officer still sitting there, waiting.

Sometime after VE day, Dad hooked a ride on a B-25 to visit a place of interest to him. During the flight, he was tired and sought a spot to nap. He curled up on a pile of tarps, slept and when they landed the pilot asked where he'd been. When Dad told him, the pilot said those tarps were laying on top of the bomb bay doors. The topper was, the bay door locks were not working and the doors were just jammed shut. Dad said his legs turn to rubber and they carried him to a jeep.

He was a lieutenant colonel at the end of the war and was promoted to full colonel during additional service in the national guard. He returned to the insurance company he worked for before the war and was promoted to several positions of responsibility. He did a great job for a man with such a limited education. As a voracious reader, he was self-taught to the point he handled his positions well.

My father did the best with what he had. My grandfather, his dad was a glass blower and hated it when he ran his thumb nail around a bottle and found a mold mark. Apparently, he cussed out those who molded glass taking jobs away from him as a glass blower. I think his occupation probably led to his death which included problems with his lungs. Dad was sixteen when his father passed, and he had to quit school and go to work to support the family. Dad had many years where most of us had a male role model; he didn't.

We were never close; I don't remember him ever hugging me. Once on a phone call, I worked up the nerve to say, "I love you." His response was, "You bet." When I was named a vice president of the savings and loan subsidiary I worked for, he said, "Young xxxxx, is a Senior Vice President over at zzzzz Bank. The "X" name was referring to a man I knew and the son of a man Dad had worked for years back. I don't know enough psychology to put a label on that approach to parenting, but it was a difficult life to live. Quite a number of other instances are popping into my mind, but I will call a halt here. I loved him because he was my father, but it was a struggle being his son.

I did manage to break one negative family trait; I not sure whether it was a one-time situation or habitual, but there was a tendency to physically strike out when anger or rage arose. After an incident one night as a child, I swore that I would never strike a woman—and I never did.

I also realize that being an introverted loaner, I found it difficult to accept and bask in the talent I did have. I would count among my faults the inability to internalize the fact that I was a hell of a lot more talented than I gave myself credit for being.

I've debated with myself whether to share this item. I carried it for a lifetime and I am weary of shouldering the load alone. Remember I grew up in Indiana, as biased and bigoted a society that existed at the time. The North was as prejudiced as the South; they were able to conceal it better. Also, keep in mind that the gay community was only mentioned in whispers; the LGBTQ+ designation was but a distant dream; and any references to being gay were done in a pejorative manner.

I was shy, I was introverted and when I tried to call a young lady for a date, I hoped no one would answer the phone. Most likely just plain teen angst, but there was no way to convince me of that. So, I went dateless for a long time. I don't remember what the ice breaker was, but I joined the homosexual dating game. I had been socializing with ladies for a short while when I got the word from my father—through a rather circuitous and surreptitious grapevine.

A man I knew was talking to my father, and mentioned that I was out on the female dating scene. Dad's response was, "That's good; I was beginning to wonder." Along with the put down, I felt his primary concern with the situation would become a reflection on him rather than me. Son of a bitch, it was hard being his son.

I did come to terms with Dad and his ways. After he passed, I told him I loved him and said goodbye to him. I think we are both comfortable with the situation; I know I am.

LOOKING FOR THE THREE-OH MARK

I have had the great good fortune to live a fantastically diverse and interesting life. I've traveled a wide range of geography over the world, staying there overnight, or a few days, or a month, or several months, or a year. They range from Canada to Thailand with a dozen other countries, or protectorates in between. Toss in a couple more foreign countries if you count just flying over them. I've lived in or traveled through forty-nine of the fifty states in America. The only state I missed was Vermont—Pat's birth state. If you toss in bodies of water like oceans, seas and gulfs, add twenty-one additional names to the list.

Traveling that much is hard; it's hard on a family as well. But in the long run, the distinct advantage we earned was a wider view of the world and our country. We have a less parochial view of events and are for the most part more well-read—if you also include television news. Too many folks here where we live were born, raised, lived and will die within fifty miles of their birth place—leaving them a bit ethnocentric.

Nearly all our moves were in the summer time, and that made it easier for our children starting a school year with all the rest. The exception was returning to the CONUS from Alaska. Their school was a couple of weeks into a new semester when we arrived. I told them the usual cliques would have formed and if they stood in the corner waiting to be accepted it would be a lonely time. I also required our daughters, the oldest of the clan to write a story about "living in Alaska." Mark was exempted since he was just beginning school.

I don't know how much help my sage advice and homework gave them as a leg up. But our girls used those stories in class and in conversations as ice-breakers. They also milked the daylights out of their Alaska tales that year and for years afterward.

For a while weather wise, we felt like Joe Btfsplk, the Li'l Abner cartoon character who walked around with a black cloud over his head. We first saw a dust devil at Big Spring, west Texas, and later viewed mile-high dust clouds—Habu's—in Phoenix that put the devils to shame. In Big Spring, the area was in at least the fifth year of a drought and had to apply for flood relieve that year. Barely arrived in Alaska and experienced a 7.0 earthquake followed by a flood that went five feet above ground level in our neighborhood. A hurricane ran up the east coast when we lived on Cape Cod and later in Houston another one passed over our house—we were in the eye. Also on the Cape, we received a 30-inch accumulation of snow. We've seen temperatures in Phoenix as high as 122 degrees above and down to minus 63 degrees in Alaska—true temp, no wind chill. That's a 185-degree swing top to bottom. Arkansas dropped an ice storm on us the first year there that plunked a twenty-four-hour electrical outage on us. A tornado raced west to east across our community about three quarters of a mile south of us. No wonder we was feelin' snake bit.

Then there was our early married life. We were susceptible to those door-to-door salesmen who plied their wares back then. Yes, we owned a Rainbow vacuum cleaner—the round one that used water in the clear plastic base as a filter. It was good, but no better than others if you changed filters as often as we dumped out that gunky water. "The Great Books" was a batch (around thirty) books of classic stories residing in their own custom bookcase. They went to family members who were better readers than we were.

And who could forget the assortment of stainless-steel cookware we purchased. The entire set survived the great Alaskan flood; several pieces were passed on to family and four of five pieces still reside in the drawer under the stove in our kitchen to this day. In life, some things you mess up and some things you get right. I think we got more of the right than we got wrong. I wish you the best of lives. My first two jobs were related to newspapers. First, I was a newspaper carrier; next I worked for the Indianapolis News, an afternoon paper, as a gopher who visited the advertising office of department stores and other advertisers with the paper and gather the ad artwork and returning it to my workplace where the artists turned into printable material.

Back in the day, many newspaper writers used a symbol to demark the "end of copy," which told the typesetters "That's all folks" for this story. One of those symbols was the number thirty (30) which was pronounced Three-Oh. So, considering my early jobs, I'll use it—here as my Three-Oh mark for this story of my life.

= 30 =

PICTURES OF THE PLANES I FLEW BEGIN ON THE NEXT PAGE. ALSO NOTE…

THE PLANES I FLEW—on page 220, I display a DVD on which I recorded videos of all the planes I flew in the USAF for twenty years. Only my family has (YT) the DVD, but I am considering uploading all those videos to YouTube. If you'd like to view them later, go to my second YouTube channel. Due to YT rules, I may need to load these videos to my primary channel. One big state of flux… I will not delay publication trying to figure out all the YT rules.

My second YT Channel – videos of the planes I flew in the USAF [or, maybe not] .
J Achor aka NOMAD - life in the air and on paper
Bitly (shortened) URL to that channel.
bit.ly/3E7lUKZ

My primary YT channel (writing and flight simulators [and maybe the "I Flew" series]) is here:
www.youtube.com/@John.Achor.Aviating.Author
Note the dots/periods between the last four words in the Handle

APPENDIX A – PLANES I FLEW

1. PA-18, Super Cub

2. T-6G, Texan

3. T-33, Shooting Star

4. F-84F, Thunderstreak

5. F-100, Super Saber

6. H-19 Chickasaw

7. SA-16 Albatross

8. B-47 Stratojet

9. KC-97G Stratotanker

10) KC-135A, Stratotanker

11) KC-135E

12) B-52 Stratofortress

13) RC-135D refueling behind a KC-135 tanker aircraft

14) RC-135S reconnaissance plane at Shemya Air Base, Alaska

15) C-47 Skytrain (military version of the Douglas DC-3)

16) & 17) The T-29 navigation trainer and VC-131 (shown here) are similar planes

18) Piper PA-28, SAC Aero Club

APPENDIX B – Stories I refer to in my autobiography –
Presentation page referenced and (App B page)

Title	Bio Ref	(App Ref)
The Golden Age of Radio and Box Tops	Page 4	(235)
Saying Goodbye	Page 5	(240)
Grace Period & A Tribute to Trous (Grace II)	Page 92	(242 & 245)
I Learned About Flying From That (PA-18 & T-33)	Page 44	(247 & 250)
Pea Ridge (a war poem)	Page 205	(254)
The Bouncing Ball	Page 119 & 140	(256)
The Grand Adventure, First Person Feline	Page 206	(260)
Callie	Page 197	(267)
Awards and Other Stuff (Saved one of the best for last)	Page 107	(270)

THE GOLDEN AGE OF RADIO...AND BOX TOPS

Do you remember when radio was king? It was not only king; radio was virtually the only in-home entertainment. Come with us to those thrilling days of yesteryear...out of the past come the thundering hoof beats of...Whoa! Great show, but not among those I really remember.

I can still hear Clayton Moore as the Lone Ranger, but those were typically self-contained episodes. The ones that really grabbed me were the continued, serial types. The names that stand out in my memory are ones like Captain Midnight...Little Orphan Annie...Tom Mix...Terry and the Pirates...and Jack Armstrong, the Allllll AMERIcan Boy! These were the heroes that drew me to the radio around four-thirty each and every afternoon. A new hero each quarter hour until 6 p.m., when something dumb—like the news—came on. Fifteen minutes was just enough time to recap the previous episode, carry the story line along for a few moments and leave us with yet another cliff hanger.

We all had a built in alarm clock even without a watch. My internal alarm was so sophisticated that it didn't matter how far from a radio I happened to be. It allowed me to reach home in time—even from the most distant point in the neighborhood. This magnificent internal device also provided for the required warm up time. Yes, we did need allow time for those vacuum tubes to come up to temperature before reception was decent.

I don't remember how many commercials they carried, but the sponsors got more than their share stuffed into each daily chapter. That may be the reason the story line advanced so slowly. Each show depended on an audience share, and what better way to insure listeners for tomorrow than a cliff hanger today.

The writers were masters of the cliff hanger ploy. Every day, day after day I was left breathless wondering how Tom or Jack or the others would get out of their latest predicament. Weekdays were bad enough, but to get me to tune in again Monday, the writers needed a spectacular cliff hanger on Friday.

I remember hearing a story about a writer on the Jack Armstrong show. He was hired on a Friday morning and would not actually begin writing until the following Monday. "Lad, how would you like to stick around for today's show?" said one of the veteran writers. "Kinda get a feel for how we work."

"That would be swell!" was the enthusiastic response.

The story line led Jack toward the end of that episode leaving him in a deep pit with no possible handholds for climbing. Beyond that, the rim of the pit was surrounded with hostile and threatening natives each armed with a razor-sharp spear.

The newly hired writer was as breathless as I was each Friday. "How in the world can you possibly rescue Jack from this impossible situation?"

The veteran writer looked at him and said, "Son, we don't share our story line with anyone not on the writing staff. Since you're new and don't officially start work until Monday, you'll just have to sweat through the weekend just like the listeners do."

The following Monday, the writer hurried to the studio to begin his first day on the Jack Armstrong writing team. With bated breath, he asked the veteran who had put him off last Friday, "How about letting me see today's script?"

The old veteran grinned and handed him several pieces of paper. "Take a look at the answer to the question you had last Friday."

Whipping past the opening credits he gazed at the first line of the Monday script. The narrator would be speaking and the line began: "After escaping from the pit, Jack…"

When all else fails, take the easy way out. I have a hunch the writers used this ploy more often than I remember. Every Monday I simply breathed a sigh of relief for Jack and the others and tuned an ear to the remainder of that day's adventure.

These types of diversions kept us tuned day after day, but just listening wasn't enough. The sponsors wanted to make money too. This is where the writers conspired with the sponsors and became masters of pushing the product. I was in my teens before I realized that Battle Creek, Michigan was not the capitol of the United States.

Battle Creek seemed the center of the universe for cereal products, which brings me to another curious question. Why did all these shows seem to have a breakfast cereal or product for a sponsor? That will have to remain an unanswered question. I never solved that riddle, yet I remember Wheaties, Instant Ralston and Ovaltine among others.

So you ask, just how did they manage to push those products? Sometimes they were subtle, and sometimes they were not so subtle. Those less skillful, were ones like Captain Midnight and Little Orphan Annie. As regular as clockwork, they would simply come up with their latest offer. Who among us could be without the latest Captain Midnight Secret Squadron decoder badge or whistle? Yes, one year the decoder came as a plastic police whistle with the decoder dial on the side. I can still picture that blue whistle and the red decoder dial.

I do remember one offer from Little Orphan Annie, an Ovaltine Mix-up Mug. Or was it a Shake-up Mug? During the first commercial, the announcer gave us the usual warning, "…stay tuned for a special offer!" Oh boy, here it comes—got to make sure I get the address correct. If not, I may miss this one altogether. A typical child's fear, no matter that the offer would be repeated for at least two weeks.

Then it struck me. Panic gripped me, my stomach was a huge knot rising in my throat and perspiration broke out on my forehead. Here I was with pencil and paper—ready to get the address…and my mother was visiting the next door neighbors. Why should that be a problem? I DON'T KNOW HOW TO WRITE YET!

I was torn. Should I stay and attempt to memorize the address or should I try to get next door and back before the next commercial. In my panic, I managed neither. My mother must have been intuitive for she arrived back home before the final commercial—when the address was again repeated—and she saved the day. Whew, talk about a close one.

On the subtle side, the real masters of marketing were the writers on the Tom Mix show. Tom would never stoop to sell anything. He cleverly left this up to his sidekick. For some reason the full name of his pal is not so clear in my mind as was the star's. Maybe that is because he was crass enough to make the sales pitch.

They all had sidekicks. Little Orphan Annie had Daddy Warbucks, or was he a mentor. Captain Midnight had Ichabod Mudd. It seems like Terry had several, but Big Stoop comes to mind. To the best of my recollection, Tom's best buddy was a sheriff by the name of Mike.

Back to crass ol' Sheriff Mike. Mike was the one who kept up a running commentary on Tom's adventures. He even explained to us dumb kids the how and why of events. Just like we couldn't figure it out for ourselves. Why, I could tell just from Mike's tone of voice that an offer was on the way.

The offer became a part of the story. Not blatantly, but very subtly woven into the plot. Ol' Mike wasn't so smart, I always knew it was coming long before he even announced the latest "be the first in your neighborhood to own the…(you can fill in the blank)!"

Tom was about to be accosted from behind by some villain as the episode ended (cliffhanger). The following day, Tom whirled in the nick of time to face the villain down (after escaping…). This time I knew it was different. Mike looked incredulously at Tom and said with a rising tone, "T-o-o-o-m, how did you know he was thar?" Boy, Ol' Mike must have thought us kids were stupid; he couldn't see an offer coming, but we could. I held my breath till the following day.

There it was. As plain as the nose on your face. Patiently, Tom detailed to the less than brilliant Mike how he knew the villain was behind him. Of course! Anyone could see that Tom was wearing a new ring. Inside the crown, was a small mirror and viewing ports. By bringing the ring up to eye level, Tom had peered into his ring and the angle of the mirror allowed him to see behind.

This time was like all others. At the first hint of a new—whatever it would be—during the next commercial break I would race to the kitchen. With a prayer on my lips, I ripped open the doors of the cupboard cabinets and searched the shelves. Since this was an offer from Tom Mix, it had to be a Ralston Purina product. Even before my frantic search ended, I knew what I would find.

Yes, there were boxes of Instant Ralston on the shelf. I may be saved yet. Oh no, all those boxes of breakfast cereal and not a single box top in sight. Everyone knows that an OFFER requires a BOX TOP as well coin of the realm. Acquiring twenty-five cents, while no easy chore in those days was child's play compared to a box top.

Racing back to the radio to make sure that they really wanted a box top this time, I wracked my brain for a new ploy of my own. "I really love Instant Ralston for breakfast" probably wouldn't work again. Evidence the numerous topless boxes already on the shelf. Maybe pepper liberally sprinkled into the boxes would give thought to weevils and my parents would throw out the old cereal. If that worked, it would be easy to request a replacement.

Those lucky kids who lived in Battle Creek. I bet they hand delivered their box tops and quarters directly to the factory and took their prize home with them. No waiting two to four weeks for the treasure. In retrospect, the time lag was really short for those days. Without a single computer to speed the process, they took half the time requested today.

At least those kids had box tops to hand deliver; and I had none! What now? I cannot remember how I was able to convince my parents that one more box top was a necessity of life; akin to food, shelter and the like. I suppose I was actually reduced to eating the cereal. Whatever the ploy, I don't think I missed a single offer in those days.

I only have one regret. I wish I still owned those mix-up mugs, decoder badges, and secret compartment rings, not to mention any number of items that glowed in the dark. Not only would they bring back fond memories, they're also extremely valuable as collector's items. What happened to them you ask? They probably were tossed out along with some 1930's Walt Disney comics—but that's another tale for another time.

The End

Previously published; "Good Old Days," magazine, September 1992. My first professional gig—they paid cash money for it

SAYING GOODBYE

John Achor, March 2007

The cold penetrated everything as the humid air of an Indiana winter can. It was mid-January 1941 and the seven-year-old boy shivered in the dark parking lot of the National Guard Armory in Indianapolis. His mother placed a comforting hand on his shoulder.

Huge U. S. Army truck engines grumbled into life and belched thick exhaust fumes. Headlights penetrated the darkness. Engines idled while men in uniform shouted orders to others and supervised the loading of the trucks. Equipment, already onboard, awaited the troops who would be traveling south to Camp Shelby near Hattiesburg, Mississippi. The boy strained to stand on tiptoe hoping for a glimpse of his father.

Hours ago, at home, his dad explained what would happen this evening. "I'll be very busy tonight, so I may not be able to talk to you. I may not even see you in the crowd. I hope so, but maybe not." The boy nodded but didn't comprehend the explanation.

He looked at the sleeve of his father's uniform shirt. The outlines of the patches were still visible. He knew they were marks of military rank. "Where did your sergeant stripes go?" he said.

"Since we are expanding to a full division, my position calls for a higher rank. I've been promoted," his father said pointing to the two silver bars on each collar. "Let's put some of these toys back under the Christmas tree, so you can play with them later."

The boy reveled in the attention his father was showing him and enjoyed the time they spent together on the living room floor.

His father continued, "I want you to know I love you and I will be thinking about you all the time I am gone. While we're alone here, I'll say goodbye now." He leaned forward and kissed his son on the cheek.

The boy shivered in the dark again. All the time I'm gone, he thought. What did it mean? He was not very close to his father, but he was used to him coming home every night. And now…Gone—where? How long? He heard the word "war" in whispered conversations between his mother and father. War. He knew the word but struggled with a full understanding of the concept. He was aware people in wars sometimes died. But…how did they die? How many? Which ones?

He was used to his father coming home each night — what if his father was one who would die in the war. A shiver wracked his body and he knew it was not from the cold.

His mother felt his body quiver and turned up the collar on his jacket. She knelt in front of him and adjusted the aviator cap and tugged the flaps down snug over his ears. "It won't be much longer," she said. "I think the trucks will be leaving soon. When we get home, I'll make you some hot Ovaltine. That will be good, won't it?"

The boy pulled the ear flaps back up—for two reasons. Only the dumb kids wore the flaps down, and the cold wasn't causing him to shiver. He looked for his father and wished with all his might it would not be the last time he ever saw him.

Pushing between some of the adults, he got a better view of the controlled chaos around the trucks. He saw a man break out of the sea of uniforms and start toward him. It was his father—his father was coming over to see him. Someone behind the man shouted, "Captain." His father stopped, turned around and disappeared back into the ocean of humanity.

His mother took him by the hand and they moved to another location—one nearer the exit gate of the assembly area. They watched and waited. At long last, the first truck bucked into motion and steered for the gate. One-by-one the other vehicles moved forward and fell in behind one another forming a long single line. The trucks reminded him of elephants he saw in a circus parade his father took him to see. He smiled.

The boy stretched to see as each of the trucks passed his vantage point. Not seeing his father, he shook his head at each truck then concentrated on the next. The last truck went through the gate, and he never saw his father.

He thought of the words: death, gone, war…the shiver wracked his body again. His mother hugged him near to her and started for their car saying, "Let's go home."

Home. The boy's thoughts turned to the hot chocolate drink his mother promised earlier. He was at peace and the images of a dead soldier melted away like the tail lights on the last truck disappearing into the darkness.

The End

Previously published, "Warrior Writers—Jouneys to Healing," by Humanities Nebraska (copyright retained).

GRACE PERIOD

Medical experts I've read say that we have a thirty-minute grace period. We'll feel no pain for a half-hour following a trauma. The concept is correct, but I'm not sure about the amount of time.

It was the 13th of January, just past the witching hour. I'm not the superstitious type, but I'm glad it was not a Friday. Small island with a military presence, western end of the Aleutian chain—dark, foreboding and known for nasty weather. Snow, sloppy runway clearing procedures, and failure to provide information to the aircrew combined to produce an inescapable circumstance. I became a passenger in a plane I was responsible for flying.

The ceiling was reasonable, but a stiff and gusty crosswind made the landing difficult. I put the bird on the runway and went through the normal procedures: speed brakes up, hand on the nose wheel steering, control column full forward and feet on the wheel brakes.

I couldn't put my finger on it, but I knew there was a problem. By midfield I committed us to the landing and shut down the two inboard engines to reduce the landing roll. Just as quickly, I knew we wouldn't be able to stop on the runway. I used the nose wheel steering to move to the right. At the bottom of the hill past the end of the runway, the terrain sloped downward from right to left. If we went off the centerline of the runway, we'd be impaled on the twin row of telephone poles that supported the approach lights. Slush, hydroplaning, runway-remaining markers flying by.

The nose wheel steering was useless as was the anti-skid braking system. The tires were not in contact with the runway; they were skimming over it on a thin sheet of slushy water. The fact that the runway was covered with slush was not transmitted to me in the air. That placed the cause of the crash on the operational staff rather than being a crew error.

I used the ailerons and successfully "bicycled" the nose toward the right. At least we'll miss the phone poles. Now, all we have to contend with is the cliff. Runway markers continued to fly by: 3000—-2000—1000—-the end of the runway.

In retrospect, I found it amazing and amusing the inane thoughts that occur in the midst of an emergency. Sliding toward the forty-foot cliff at the end of the runway, in a sixty-million-dollar, one-of-a-kind airplane, I noticed that we would run over and break a couple of $50 runway marker lights.

Just prior to the end of the runway, I cut the two remaining engines and rang the alarm bell—one long sustained ring—prepare for impact. As the engines wound down, the generators tripped and the lights went out. We hurtled off the cliff into the dark void that surrounded us. Since it was pitch black, I didn't brace for the impact; I never saw it coming.

Crunch—crunch—-and a final grinding screech. The gut-wrenching impact ripped landing gear and one engine from the airframe, and it broke her back. The fuselage had a gap large enough for at least one crew member to use as an emergency exit.

After we scraped to a halt, I released my seat belt and shoulder harness and looked down for the flashlight I kept on the cockpit floor near my right boot. It was easy to find. The impact was kind enough to turn it on for me.

I started for the rear of the aircraft. I needed to assure myself that everyone was out. I swung the beam of light from side-to-side, calling out—all the while thinking: this son-of-a-bitch is going to blow up any minute. We had landed with nearly eight thousand gallons of fuel onboard. I could picture it spilling and leaking from broken lines, broken tanks, and seeping toward some hot piece of metal.

I saw no one during this trip through hell. At the aft hatch, I found the escape rope dangling through the open hatch. I'd left my copilot up front struggling to free a boot pinned by twisted metal. I dreaded the trip back through that dark tunnel; I was still sure the damn thing would erupt into a flaming inferno. But again I needed to know whether he had escaped.

Saved. Through that open hatch, I heard my copilot calling to me from outside. I grabbed the rope and slowly slid down the fifteen feet to the ground.

My crew was out. Two other team leaders confirmed to me that they had a full head count. All eighteen of us survived and were huddling in the chilly night air, upwind from the wreckage. I zipped up my light weight flight jacket, lit a cigarette, and nearly collapsed. My legs sagged and it took two people to support me. They helped me into a vehicle for the ride to the infirmary. Verdict: severe lower back sprain from whiplash. It was a week before I could get out of bed without sliding from the bed on my belly, putting my knees on the floor and then lifting my body with my arms.

I did have a grace period of pain-free activity. One long enough to do what I felt I had to do. My best guess is that it was less than half the allotted time the experts forecast. But it was long enough.

The airplane? They salvaged one of the four engines and several thousand pounds of electronics. They hauled the carcass to the base dump. All that damage, but no fire, no explosion. Eighteen walked away with only a few minor injuries. Well, seventeen walked and I was carried.

That model plane had a jump-seat that was a folding affair and poorly anchored. Since it wasn't stressed for impacts or accidents, it was reserved for instructors and evaluators. In the past, I'd stretched the regulations about other crew members occupying the seat on landing. That night, no one asked. Had they asked, I may well have allowed them to ride the jump-seat for landing. And if I had, I have no doubt they would have been seriously injured or killed. Though it was empty and stowed - I've always believed that someone was riding the jump seat that night.

The End

I tell the story in detail in my memoir, "The Iron Pumpkin." Check it out on Amazon.

Previous contest entry

A TRIBUTE TO TROUS, GRACE PERIOD TWO

If you've read "Grace Period," you know it is the story of a major aircraft accident in January 1969 and discusses how long adrenaline can stave off pain. I won't recount all those details again—only enough so this part of the story makes sense.

A bit about my background may help us both understand why I am putting this down on paper. While you might judge me as—at worst, an agnostic and at best, a deist; I also believe what Ernie Pyle, the World War II war correspondent, said on the subject: There are no atheists in foxholes. I've been in a number of foxholes, more figurative than literal, and I have called on the Almighty on those occasions. That's the dichotomy I faced accepting the conclusion I reach at the end of this tale.

I flew with Ralph W. Trousdale; he preferred to be called "Trous, as in trousers," in the mid-1960s. He came to Fairchild Air Force Base in Spokane, Washington under a "cloud." Trous transferred there from the 1st CEG (Combat Evaluation Group) at Barksdale AFB, Louisiana. The CEG are the grand gurus who told all us line pilots how to fly the KC-135 (a four-engine jet, air refueling tanker).

Trous and his CEG team visited Fairchild earlier and ripped the knickers of many a pilot in our squadron. So, when he arrived as just another line pilot, it was payback time.

He could have skipped his check out flight at Fairchild. The grade was a foregone conclusion—FAILED. I was assigned to his crew following that initial check ride and we were under the gun. We all passed his second flight check and went to work as a crew. I soon learned that Trous had forgotten more about the KC-135 than the rest of us would ever know about the bird.

As luck would have it, CEG visited Fairchild again, and this time failed several Standboard crews including the pilots on the senior crew. Standboard crews were the local gurus of flying and the equivalent of a local CEG. The most experienced Aircraft Commander in the squadron, and the most capable of assuming the empty Standboard position was - you guessed it, Trous. After only a few months of flying with him, I found myself the senior Standboard Copilot.

I learned more about life and about flying from Trous than I'll ever be able to remember. He was always a teacher, even when I didn't realize it. Trous was a true mentor, and I will forever be grateful to him for what he imparted to me. My only regret is that I did not tell him this before his fatal heart attack.

I upgraded to Aircraft Commander and had my own crew. Later, I was transferred to Alaska to assume command of a reconnaissance crew. The last I knew, Trous was still at Fairchild. We lost track of one another over the years.

It's time to get back to the aircraft accident. On that cold January night, I was flying an RC-135S (a strategic reconnaissance version of the KC-135) from Shemya Air Base, Alaska. They've changed the name of the base, but the island is still way out there toward the Russian end of the Aleutian Islands.

About thirty minutes past midnight, I landed on what was later determined to be a slush-covered runway. I didn't know it, but our fate was sealed early on landing. I cut engines and tried to slow the aircraft. When I became aware that we would not stop on the runway, I did my best to move the airplane to the right side. The nose wheel steering was as useless as the wheel brakes due to the slush. I used the ailerons to "bicycle" the plane to the right. We needed to go there; otherwise, we would be faced with a twin row of telephone poles supporting the landing lights for the opposite approach.

We made it, went off the right corner of the runway and over a forty-foot cliff. All eighteen of us escaped with a few minor injuries. I told the accident investigation board I remembered twisting the nose steering wheel to the stops with no effect, and I remembered using the ailerons to do the job. In "Grace Period," I said I always thought someone was riding the jump seat that night. That's where the story rested for all these years.

Several months ago, I woke up in the middle of the night and remembered something more. It was Trous who taught me how to use the ailerons to "bicycle" an airplane. I am firmly convinced that if we had gone off the centerline of the runway, the three of us in the cockpit, my Copilot, one of my Navigators and I, would not have survived being impaled on those phone poles. The destruction those poles would have caused the plane, might have affected those who were further aft as well.

I'm not sure if I believe in guardian angels or not, but now I think I know who was riding the empty jump seat that night. Thanks, Trous.

The End

Previous contest entry

I LEARNED ABOUT FLYING FROM THAT

THE PIPER CUB — PA-18

Previously published in *Computer Pilot* magazine ca 2009

This is the first in a continuing series of articles. While I plan to take a lighter approach with these stories, the "lessons" are meaningful and serious. For my initial topic, what is more appropriate than the first airplane I flew?

As a member of the last class in the U.S. Air Force (57-I, 1956) to train in tail-draggers, we began our flying careers in the Piper, PA-18 Super Cub. All in my class claimed that anyone could land tricycle gear aircraft, but it took a real pilot to put the tail wheel down first; at or near stall speed. In reality, only one third of the class 57-I flew the PA-18. My class was large enough to be split between three Primary Training locations. The other two bases had already transitioned on to the next generation of aircraft for flight instruction.

Super Cub was a fancy name for a J-3 Cub with a Lycoming 4-cylinder, twin opposed 108 horsepower engine. The plane is still around; I found several PA-18s for sale on the internet, ranging in price from $40,000 to over $100,000 (USD). Although I can't remember using flaps, the internet tells me the Super Cub was equipped with them. I also do not recall a radio, so in the air it was see and be seen. The instrument panel was little more than needle, ball and airspeed plus a Whiskey compass, tachometer and altimeter — but did not include an attitude indicator. After all, this was a VFR airplane and the makers figured a pilot could keep the wings level and tell up from down without one. One instructor described the turn and slip indicator saying the needle tells you which way you are turning and the ball is used to make the turn coordinated. Then he added, don't be too concerned about the ball, as long as you keep it inside the cockpit, you'll be safe.

Beyond that, it's a high-wing, yellow fabric stretched over a steel tube frame, a tail wheel and just enough room to accommodate a student in the front seat and the instructor in the rear tandem seat. When taxiing a tail-dragger, good safety protocol—and our instructors—required making S-turns down the taxiway to avoid people and the whirling props on other planes. We never made wide enough turns according to the instructor. Sitting behind and lower than the student, he couldn't see as far as we could. Even though making the turns was a pain in the tush, it beat bending the propeller on a solid object. This lack of visibility also created problems for maintaining a straight line for takeoff. Straight ahead progress could be monitored by comparing the amount of runway visible on each side of the nose. This was a continuing process until the tail flew off the ground and the pilot could see over the nose.

I was stationed at Bartow Air Base, Florida (USA) which was about half way between its namesake city and Winter Haven where I lived. Bartow AB was our "home," but along with our North American T-6G Texan planes, it was becoming crowded with Beechcraft T-34 Mentors and North American T-28 Trojans which would serve as the training aircraft for following classes. We were shunted off to an auxiliary field nearby. Not much there beyond a briefing building, a small parking ramp and a couple of runways. Near the shack, was an open livestock watering tank about two-feet deep and around twelve feet in diameter. That's where the ritualistic "dunking" came for those of us who soloed the yellow Bamboo Bomber.

We got around 20 hours in the Cub and near the end of our training in this plane, one of the instructors flew a T-6G into our auxiliary field so we could look it over. I remember wondering if I would ever be able to herd such a huge plane around the skies. It was way larger than our Cubs but compared to four-engine jets I flew later; it wasn't all that big. The Cub offered a low cost airframe and low operating costs compared to other trainers. It was a great platform for teaching students the fundamentals: takeoffs, turns, power-on and power-off stalls, forced landings, and the all-important step—getting the bird back on the runway all in one piece. And that last one was the basis for learning a lesson that day.

This was Florida, with scrub vegetation and bunches of sand in between. Off the approach end of one of the runways was a circle of sand that must have been 50 feet in diameter. On a hot afternoon, I turned final with my instructor and set up my typical angle of descent to the runway. As I passed over the patch of sand, the thermal shoved my little Cup upward—for what seemed like 500 feet. I exaggerate; it was probably only ten or twenty feet, but it messed up my final approach. I managed to get the nose down and put the plane on the ground. After a full stop landing, we taxied back and waited in line for another takeoff and a second pattern.

On crosswind, downwind and base legs, I plotted how I could defeat the sand pit. This time around the pattern, I figured I'd get ahead of that updraft just before the threshold. I turned final and started a steeper descent with the idea the thermal would bring me up to the glide slope and I could impress my instructor with a perfect landing. Ah, the best laid plans of…

I sank lower and lower on final waiting for the thermal to save my approach. No such luck. At last as I added a bit of power to keep from going dangerously low—here comes the updraft. With the extra power I'd added and the thermal, I'm back where I was on the first approach. Scrambling with the controls and power, I managed to squeeze out a barely acceptable landing.

I suppose the lesson is: don't mess with Mother Nature. Rather than trying to outguess what the weather will do, I would have been better off flying the airplane, keeping the glide slope constant, and being prepared for odd eventualities. I could have counteracted the updraft with forward stick pressure if I had been prepared.

This was no doubt, not the day my instructor cleared me to solo. Eventually, he did send me into the wild blue yonder all by my lonesome and I was dutifully dunked in the cattle trough and prepared to move on to the T-6G Texan.

I've always tried to tuck lessons learned away into the back of my mind and carry their value with me. Stay ahead of the airplane; if you are continually trying to catch up with the bird, someday—something will reach out and bite your backside.

Keep in mind that it's not one horrendous event that will cause the accident; it is more often than not a series of smaller events that accumulate and pile up. Use all the lessons you learn about flying, plan ahead and don't find yourself at the confluence of all those small events.

The End

I LEARNED ABOUT FLYING FROM THAT...
The T-33 Shooting Star at Webb AFB, Texas
Previously published in *Computer Pilot* magazine ca 2009

For the fourth in this series, I've left primary training behind and moved on to the Basic Flying Training program at Webb Air Force Base, located at Big Spring, Texas (USA). We flew the Lockheed T-33A Shooting Star, a single engine jet with a pair of tandem seats. The image of the airborne T-33 is from FS2004.

T-33A airborne near sunset

Perhaps Shooting Star was a bit of a misnomer for the T-Bird because this plane was equipped with an Allison J-33 centrifugal flow jet engine. That meant a large turbine wheel spinning perpendicular to the air flow shoved air into the engine. Those engines lost favor early and nearly all engine designers went to axial flow engines, where the air was forced in more of a straight line into the combustion section of the engine.

John entering the cockpit – posed since we never flew in Class B, shirt, trouser and tie

The drawback to the centrifugal flow engine was an inherent delay in acceleration. You could throttle-burst the old gal from idle to 100%, and then take a short nap waiting for the engine to wind up to maximum power—it actually took only 12 seconds or so, but that can be a lifetime in an emergency. That was our first lesson in this airplane. Carry a minimum of 50% throttle and don't get behind the power curve, because the time needed waiting for full power could run you out of airspeed, altitude, and ideas all at the same time. This is a picture of me climbing into the cockpit of a T-33 for our class "publicity" photo.

Forced landings were always "fun." The instructor would yank the throttle to idle and announce "forced landing." We tried for an overhead 360-degree pattern and knew if we hit high-key around 8,000 MSL, which would be about 5,500 AGL at Big Spring, we could put it on the runway. To maintain visual contact with the field, high-key was a point adjacent to the landing end of the runway and on runway heading. Low-key was a point on "downwind" where the pilot decided it was time to begin the turn to final approach.

This image is from the T-33A Dash 1 (Technical Order: T.O. 1 T-33A-1).

Mastering that one, the next step was a no-flap forced landing. High-key was the same, but this time we actually rolled out on a downwind of sorts. As my instructor described it, "Hold downwind (low-key) until you are absolutely sure you can't make it back to the runway and then fly another thirty seconds."

That was close to the truth. A clean T-Bird would glide forever with nothing hanging to slow it down. One of our classmates sheared a major shaft in the accessory section of the engine, leaving him with airspeed, altimeter, stick control (with unboosted ailerons), about thirty seconds of battery power and no landing gear, speed brakes or flaps. He was able to try a restart and call the base declaring an emergency before the battery died. He departed, Lake J. B. Thomas, northeast of the base at around 20,000 feet, knowing he could glide the twenty-five to thirty miles and make high-key at Webb with no sweat.

Big problem looming; he had never even seen a no-flap forced landing let alone fly one. He overflew the approach end of the runway at the correct altitude and began his pattern. He came over the touchdown point a little hot and floated past 10,000 feet of runway and finally touched down a half-mile or so past the concrete—big cloud of dust in the boonies. After debriefing, he came back to the ready-room and said the instructors wanted him to sign a backdated training slip showing that he had been demonstrated a no-flap forced landing. We all advised: Don't Sign. The fact the instructors didn't have a training slip to that effect was his only defense against being pressured out of the program—called SIE (Self Initiated Elimination). That could happen fast. One day a student was there and gone the next. The answer to whispered questions, was usually; SIE. He didn't sign; he graduated on schedule.

Like the example above, if we were lucky and paid attention we learned from the successes and failures of others. Inattention to flying or preparation for flight was one of the first things that could kill you. I've been fortunate because I haven't lost many squadron mates and friends. At

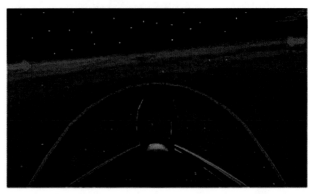

least a couple of dozen seems a small number compared to all those I knew and flew with. I hope that doesn't sound callous; being a pilot for twenty years produces a rather pragmatic approach to life and how quickly it can end while flying. Ernest Gann, wrote "Fate Is the Hunter" and in it a main character – Rod Taylor as Captain Jack Savage (how's that for the name of a swaggering heroic pilot) answers his copilot who is worried about weather, low fuel and the doubtful location of the runway. Savage says (as best as I remember the dialog): It can't not be there. Otherwise, we'd crash and it isn't time for that.

Before you read this section, take a look at this cockpit view at night. When you've determined the attitude of the aircraft, press on.

Late in the training program, I needed thirty or forty minutes of night flying to fill the last squares for graduation. My usual instructor was tied up, so I drew a little fellow using Rooster as a call sign. We cranked up, took off and climbed to altitude. Since all I needed was the flying time, we were left to our own devices as to what we would accomplish.

We did a bit of sightseeing and worked our way west of the base. That part of west Texas is sparsely populated. After a while, Rooster asked if I'd ever done unusual attitudes at night, and I said, "No." Unusual Attitude Recoveries were a routine practice exercise, but typically done during the daylight hours. The student would lower his head and close his eyes, while the instructor maneuvered the plane into an unexpected situation and attitude. When he was ready, the instructor would call "recover" over the intercom and the student checked all visible references and brought the plane back to a straight and level condition.

Rooster gave me the usual instructions and I dutifully lowered my head and closed my eyes. He was smooth with the stick and I could feel several turns and changes in attitude. When I opened my eyes, I saw roughly what I captured in the accompanying screenshot. I thought, this one is pretty easy, I figured he would really try to screw me up. It looked to me like a nearly wings level attitude with the nose slightly below the horizon (see red arrows in the image pointing out the horizon). With a nose low recovery, the first step would be to retard the throttle to prevent an increase in airspeed.

About that time, I noticed the airspeed was decreasing! Oops; bad first evaluation. Around this same time, I began to feel myself hanging in the shoulder harness. A more complete scan of the instruments gave me the correct picture. I was nearly wings-level, but we were nose up and inverted. Once certain of my attitude, recovery was simple and the instructor gave me a verbal pat-a-back. Lesson—make sure you thoroughly evaluate your situation before applying corrections. Using my first guess, pulling back on the stick to "raise" the nose, we might well have ended up screaming straight down toward Mother Earth.

Next time, I'll see what other Shooting Star events come to mind.

The End

PEA RIDGE – A WAR POEM

John Achor

In 1998 before our move from Arizona to Arkansas, Pat and I took a survey trip. Driving from Hot Springs northward toward Fayetteville, we saw signs announcing the Pea Ridge National Park, a Civil War battlefield and decided to interrupt our trip.

We drove the perimeter of Pea Ridge and walked over part of the grounds. As I did, the idea for this poem struck me. The first two lines...

> I hear the crack of musketry, an' I hear the cannon's roar,
> I never figured that fightin' was, such an awful chore;

...remained virtually the same as those in the final version – came to me there on the battlefield.

Four years later, I finished the poem.

Pea Ridge

Arkansas, Civil War Battle

I hear the crack of musketry, an' I hear the cannon's roar,
I never figured that fightin' was, such an awful chore;
Up a country road, like me an' Pete as kids would walk,
To get to them fishin' cricks, or jes to sit and talk.

I'll write to you my darlin', as soon's the day is done,
When there's some time to spend, when the battle's won;
I'll tell you how ol' Pete an' me, has come along this way,
An' hope and pray that we live to see, jes one more day.

The skirmish line's a formin', an' we're movin' through the trees,
The wind is kickin' up a bit, it's a cool an' chillin' breeze;
It'll dry them stains of sweat, an hide them signs of fear,
I dare not let my friends see, much less the brigadier.

The firin's growin' louder now, I can hear the Minnies fly,
Whippin' through the twigs an' leaves, like death's mournful cry;
We're on the move agin', it ain't much after dawn,
Some of us is fallin', an' some is marchin' on.

The Yank raised up and aimed his gun, at me or was it Pete,
I shoved my friend aside, lookin' for cover in the battle's heat;
But 'fore I'd moved I'm down, I'm down before I'd fired a shot
Oh Lord that ball it took me down, my belly's burnin' hot,

I'm lookin' down now, on a field of crowded stones,
Holes in the ground, that are filled with guts and bones;
There's a stone I see, a sittin' atop of fresh turned sod,
It's the one for me, but says: He's known only to God.

It's me I cried, it's me, it's me, it's me,
Don't cha know who I am? Please come and see;
Then I seen the stone next to mine, an' it was for ol' Petey,
Ol' Pete'll never let me down I thought, he'll remember me.

The long day's done, it's quiet now, an' the fightin's o're,
Twilight's come and we can rest, don't need ta march no more,
I'm sorry my darlin', I won't be comin' home, no steps must I retrace,
Here I'll rest for an eternity, just waiting to see your pretty face.

© 2001 John Achor

Previous contest entry

FOLLOW THE BOUNCING BALL (RIVET BALL THAT IS...)

An article by DC-3 Airways (DCA-324) Technical Editor, John Achor

When you finish this page, if you would like more information about Rivet Ball and the companion plane, Rivet Amber (lost in June 1969) and the men who flew on them, you can use this link to a site by King Hawes (rc135.com).

THE ALASKAN RING OF FIRE CHARTER

I just finished flying "The Alaskan Ring of Fire" charter created by Bob Reid, DC3-375. Great flying—interesting navigation problems—short fields—and some surprises.

There is a tall radio antenna to the right side of the final approach at the last destination—did anyone notice it, or were you distracted? I'll leave it at that 'cause I don't want to spoil the fun. Good work Bob.

REAL LIFE IN THE ALEUTIAN ISLANDS

Here's a bit about the actual flying I did out in the Aleutians. We flew an RC-135S (Rivet Ball) aircraft from Eareckson Air Station (AS). Back in the late 1960s, it was called Shemya AS. I've included a few pictures from the island and a couple of screen shots from the charter flight.

A bit about the runway at Shemya. There was a single 10,000 foo t runway (28 / 10). At the west end of the runway, there was about ten feet of dirt "overrun" before reaching a 45-foot cliff. Extending out on the runway centerline was a twin row of telephone poles—holding the runway approach lights for a landing on Rwy 10.

The weather at Shemya was generally miserable. I've seen 25 knot winds and zero visibility in fog for more than a half hour. So, you say, you can't have fog with that much wind—you can if the clouds are on the deck and the cloud bank is large enough. I'll say one good thing about the weather. The accident review board was delayed arriving at the island for a week due to winds and weather. The day they arrived, as they stood on the ramp, the weather went from CAVU (clear and visibility unlimited) to zero-zero conditions in blowing snow and then back to CAVU—all in thirty minutes.

The night of the accident, I was flying a GCA (Ground Controlled Approach, precision radar) approach to Rwy 28. The reported Runway Condition Reading was RCR-09IR-P. The IR stood for icy runway and the P indicated patchy, but the 09 reading was well within our safety margins for landing. What they didn't tell me was that they had sprayed isopropyl alcohol on the ice. That turned the ice to slush and I hydroplaned off the end of the runway and over the cliff. After the crash, the RCR was confirmed as less than 04 and that would have sent us to our diversion base (Eielson AFB, Alaska).

The nose steering was totally useless. The anti-skid system worked like a charm— only one trouble—the tires were not in contact with the runway surface. I managed to "steer" the airplane using ailerons to "bicycle" us toward the right edge of the runway and avoid those telephone poles.

The airplane was a virtual total loss. They salvaged one of the four jet engines, part of the tail section and about 60,000 pounds of electronic gear from the interior. Staff operational errors and weather were listed as the causes of the accident. No responsibility was assigned to any crew member.

RC-135S (Rivet Ball) departing Majors Field (east of Dallas, TX) after undergoing modifications at LTV (Ling-Temco-Vought) USAF photo

The Rock seen in real life in the winter time

Shemya (The Rock) seen from Flight Simulator 2002

AND THEN THERE WAS JANUARY 13, 1969

A look at Rivet Ball sitting at the bottom of the 45' cliff on the day after… one of the guys escaped through the crack in the fuselage at the left-wing root…

ALL 18 CREW MEMBERS WALKED AWAY
USAF photo

Final approach to RWY 10 flying the venerable DC-3 (Gooney Bird) for DC3 Airways (a virtual airline).

Note the wreckage of Rivet Ball in the lower left of the image (Photoshop of a screenshot)

During the charter I ignored the ILS to Rwy 28, entered downwind for Rwy 10 and here I am on final (in the photo above). If you look closely, toward the bottom left, you'll see the remains of the RC-135S sitting in the "Million Dollar Dump" on the west end of the island. Don't look for the wreckage in your version of Flight Simulator, this one only appears on my computer. An examination of the wreckage will tell you I did a copy / paste to the original screen shot :-)

Here's an actual photo of what was left of Rivet Ball after the salvage effort

Photo by G. Smith

The End

© 2003 John Achor

Previously published on the internet, DC3 Airways (Virtual Airline).

THE GRAND ADVENTURE – FIRST PERSON FELINE FROM MY POINT OF VIEW

© 2002 John Achor

Translated from a web format.

This story is an award winner and took top honors (without the benefit of the photos) in the Board of Directors Non-fiction category at the Ozark Creative Writers in Eureka Springs, Arkansas, and put $100 in my pocket.

My name is Lexus and I've decided to tell you the story of how my People uprooted me, and dragged me across half the country in what they call…The Grand Adventure.

I suppose I should introduce myself and my People. I'm twenty-eight—Oh! That's about four in people years—and my People are Pat and John. They're both around nine and a half—Oh, again! That's cat years. John and Pat Person are both pretty decent…at least they feed me on a reasonable schedule and toss in back rubs and brushings on a regular basis. I'm a loveable female with an extremely sleek, luxurious coat. It's no wonder those auto folks named that car after me. I found Phoenix, Arizona a fine place to live. Since I'm an indoor type personality, what do I care that temperatures can hover above 110 degrees for weeks at a time.

I had a premonition of impending doom when the boxes began to pile up. The only way I could keep a rational handle on the situation was to ascend this ladder thing and observe the happenings from my aerie. "My," you say, "what an extensive command of the language." Just look at the picture above, and you'll see that I do study the dictionary quite a bit. I've learned quite a bit from that tome, albeit via osmosis.

Perhaps I should digress long enough to let you know that I am writing this missive. My command of English is far beyond what one might expect of a cat. It's my "cat" that's a bit lacking. You see, I came to my People from a Rescue group, and I was separated from my mother and siblings, as well as from other felines, at a rather early age. I had little time to be socialized into the world of cats, so the language you would normally expect from me consists mainly of chirps, purrs, and an occasional chin quiver which is a cross between a chirp and a purr. Those only occur when there's a bird, or at least a very large insect, in sight.

I also should warn you. Though, my language skills are awesome, I do have trouble with the time and space continuum. I also flunked geography. When Pat and John Person mentioned Arkansas, I had no idea where it was or just how long it would take to get there.

Yes, I am writing this, however, I've asked John Person to do the typing. I can type, but keyboards were designed by someone who had no comprehension of feline anatomy. My paws tend to strike two or three keys at the same time. I can read it, but you folks might have a problem with it.

Back to the Grand Adventure. My People began moving furniture out of the house—said it had to do with shipping weight and the fact that this was the first long distance move they would have to make totally at their own expense. Imagine, stacks of drawers were all that was left of a huge chest and triple dresser just before a nice young man hauled them away.

Furniture, rugs, all sorts of belongings began to disappear. The only stable items in my environment were my food dish, some carpet remnants—that I commandeered after the treadmill went out the front door—my litter box and of course, MY Afghan.

The fateful day loomed. I could tell even though Pat and John Person tried to keep if from me. They had purchased a new pet taxi. The old one, a hard shell plastic, didn't hold many—if any-pleasant memories. It was generally used to lug me to the vet's office or to the pet resort. The resort wasn't bad, but that vet! You can't imagine the indignities I suffered at the hands of those doctors. I mean, shots are bad enough, but when they want to take your temperature—ugh. I'll leave that to your imaginations.

Speaking of the vet, during our last visit he gave my People something called tranquilizers. He described them and said they would keep me calm, how to pop the pill down my throat—not likely—and that I should not be fed afterwards. Huh! Not likely either. Well, whatever they were, John Person decided to disguise one by pressing it into a couple of soft treats. The munchies were great and I really didn't pay much attention to that hard center. Small problem, I barfed yellow shortly after I downed the tranq laced treat. I managed to perform in a similar manner the second time John Person slipped me one of those things. That John Person tries hard, but a second clean-up cured him—no tranqs for the trip. I'm pretty mellow and I really didn't need them. Glad he didn't wait till we were in the car to test them. Can you imagine what he'd have said about yellow spit-up on his car seats.

They only used the new carrier for pleasant trips. John Person would mutter something about going out for yogurt, and off we'd go in the car. Them in the front seat and me in the back seat. They opened the top of my new taxi, but I'm too smart for them. Mostly I just stayed inside the carrier, though occasionally I ventured out onto the back seat. I could look out the windows while they sat there stuffing their faces with soft ice cream. When John Person said "we" were going for yogurt, it must have been the royal "we" pronoun. I don't remember getting any.

If you look real close at the picture of me and my pet taxi, you can see that blue thing my People call a harness. They put it on me even before we started our adventure. They said it was so I could get used to it. Like that's going to happen in this lifetime. Anyway, every time they open a door—in the car or in a motel—one of them grabs me by that harness. Sometimes they even hooked a leash to the harness. Said it was so I wouldn't run away. Like that's a possibility—I'm an indoor person. "Better safe than sorry," my People said. Whatever keeps them happy.

It was happening. I was shoved into a bathroom along with food and my litter box. When they let me out, the house was completely empty. I had a sinking feeling and sure enough, into the car we went that afternoon. It was a short trip, we only ventured to the other side of town. My People said something about getting a jump on traffic the next morning, but I could tell they just wanted to test my staying ability in the back seat of a car and at a new house.

It looked like a house, but it sure was small; they referred to it as a motel. I explored this new abode from top to bottom and it was okay. No Taj Mahal, but it would due for a new home, and the bathroom floor had some really cool tile which was enjoyable.

Bright and early the next morning we were off again. Another car trip, and it didn't look like we were going to return to that new house—all my belongings were tossed into the back seat along with me in my carrier. Was I doomed to spend the rest of my life riding all day in a car to get to a new and different home every night. Maybe, maybe the next day would not be a repeat of this tiresome ritual.

Boy, was I wrong. We spent another entire day in the car. By now, I was concerned about Pat Person. She was nowhere in sight. John Person kept picking up this black rectangle thing—he called it a Cee-Bee—and he talked to it. The next thing I know, Pat Person is talking—I know because I can hear her voice—but she's still nowhere in sight. She must have been around somewhere; she kept appearing each time we stopped and each night in the new houses.

Finally, I ventured out of my taxi and explored the back seat. By now I was in need of my litter box. It was not to be found. I did locate a funny looking container with some gravel in it on the floor of the back seat. It made a fine place to nap—took my mind off my bladder for a little while. I've heard that some seats in a car are more dangerous than others. But as any intelligent cat knows, the safest location is hunkered down on the floor in the back seat. Well, John Person got upset when he found me snoozing in the gravel. He put MY Afghan on the back seat floor, and he put that container thing up on the back seat. He tried to coax me into it by playing with the gravel. No thanks.

I still needed a litter box. I explored the car and climbed into the front seat where I found a bit of food and water in my dishes on the floor. No litter box. I hated to do it, but when nature says that's all she wrote, that's all she wrote. I hopped into the back seat again leaving a small wet spot on the front floor mat. To say that John Person was not a happy camper would be an understatement. I couldn't believe he'd forgotten to bring a litter box. The next thing I know, he's wiping up my piddle-puddle with an old T-shirt rag and then he puts that in my new gravel snooze box. Oh! A lightning bolt of realization. Could he be trying to tell me that this is a litter box? Well, if it is, why did he wait till we're on the road to introduce me to it? It doesn't look like a litter box; mine is a hard shell with a cover and filled with sand. This new thing is soft plastic, a different shape, a different color, no lid, and there's only this gravel stuff in it.

From his gyrations, I finally figure John Person does want me to use this thing as a litter box. Okay, whiz, whiz. Then that night, John Person cleans out this new litter thing. I mean he really gets it clean. Next day, I had trouble telling it was the correct spot to deposit my waste products. I tell you, that boy tries hard, but sometimes he's a couple cans short of a six-pack.

The day ends and we go to another home. Not any bigger than the one we just left. It must be one of those motel things again. Here we meet my uncle. I don't know why they call him an uncle, he's their son. I explore this new house and guess what? They put the beds on top of a wooden box—and if you're real clever you can locate the entrance. From under there, I can hear my People and my uncle looking all over the room for me. I just smile and hunker down in my nice dark hiding place. Ha! If they can't find me, we'll never have to leave this new house.

Bummer. The next thing I know, John Person and Uncle Mark are lifting the bed up off the wood frame. Pat Person pulls me out and they stuff the bed spreads around the end and sides of the bed. I can't find any doorway into my new found hiding place. Occasionally, my People are smarter than they look.

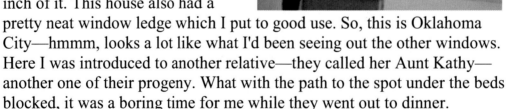

We packed up for what I assumed would be another day in the car—right again. After what seemed to be an interminable period of time we arrived at still another of those motel-homes. I know it was different, 'cause I checked out every inch of it. This house also had a pretty neat window ledge which I put to good use. So, this is Oklahoma City—hmmm, looks a lot like what I'd been seeing out the other windows. Here I was introduced to another relative—they called her Aunt Kathy—another one of their progeny. What with the path to the spot under the beds blocked, it was a boring time for me while they went out to dinner.

Next morning, same drill. I did perk up when I heard my People say something about this being our last day on the road; it's about time folks. The day started off bad—drizzling rain as we loaded the cars—and got worse. According to John Person, we had seven hours of constant rain. Sometimes so hard he could barely see the road. In fact he said, "In fifty years of driving, I've never seen storms this bad." I guess that's bad, how long is fifty years?

We turned south off the interstate and the sun came out. We had great weather all the way into Hot Springs Village. At last a good omen and a permanent place to stay.

This one really was a house; it had real furniture. No more of this one room stuff. Permanent, Ha! Two nights in this place, and then we moved again. Thank heavens this time it only took a few minutes. I'm getting settled into this really permanent home. My People put my carpet remnants down on the floor for me and MY Afghan is on a window seat where I can get some sun and enjoy the view. Got to tell you, this Arkansas place is more interesting than Arizona. The bugs are bigger and just the other day a furry little critter with a bushy tail came up on the deck so I could scope him out.

There are lots of places to explore, like the tops of the book cases in the living room and the closet where Pat Person keeps her clothes. Then there's a new place like I've never seen before.

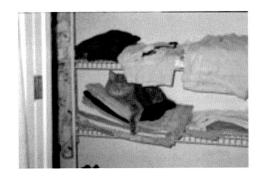

They call it the over-the-John storage place. Wow, that boy even has places in a house named after him. Sure is a neat spot for a kitty to sack out and grab a nap.

I'm glad we made the trip, but I sure am happy that we won't be doing any more traveling. This new place is very nice, and my People said we're here to stay. They mumbled something about building a new house. I'm not sure what that entails, but we're in this permanent home for six months. Guess that means no more car trips.

Can anyone out there tell me what six months means?

The End

Previous contest entry and First Place winner.

GOODBYE, CALLIE

Callie was not the most original name for a small, female Calico cat, but it suited her. She did her best to live up to the Calico reputation, she was independent, curious with a mind of her own. She was also intense and sometimes, downright ornery. My wife rescued her from the Humane Society's shelter.

She did not care for the firm voiced CALLIE that I reserved for those disciplinary times. Most of the time she ignored it. Perhaps it was the behind-the-hand smile on my face. I think she knew just how amusing we found her even when she was being obnoxious.

She was not with us long. Callie was barely three years old when she died of a massive heart attack. I'm seldom gone in the evening, but I was out the night my wife found Callie dead when she arrived home from work.

I felt terrible. That morning, I was aware that Callie wasn't feeling well. She'd been less than perky, so a few days earlier; the vet examined her and found nothing specific. He gave her a shot of antibiotics and told us to bring her back if there was no improvement.

I broke our rituals and rules that morning and gave her a treat because she looked like she needed something to lift her spirits. She didn't eat them before I left and that was unusual. I knelt next to her and stroked her head. I'm glad I took those few extra seconds with her. The only positive aspect about it was that sometime during the day, she ate the treats I left. I can hope when she consumed them, she remembered some happier time in her life before her heart gave out that day.

Following a necropsy, the vet told us that the inner walls of her heart were thickened and caused it to work too hard. There was no enlargement of the heart that would have been visible on X-rays, and timely diagnosis was probably out of the question. We went to bed about the usual time. It was nearly midnight when I arose and knew I had to do something. I needed to say goodbye to Callie. I didn't want to see her dishes, toys and other belongings around in the morning and also knew that my wife would appreciate not seeing the reminders.

Downstairs, I busied myself gathering her possessions. I knew that we would eventually get another cat, so I had no thought of throwing everything away. I'd simply pack her things away for a happier day. I loaded everything into large bags and stored them in the garage. I even packed her treats, but for some reason stuck several in my pocket.

As I picked up everything, I found a few strands of her fur. The trash bin was not handy, so I tucked them into my shirt pocket. Then there was only one more trip left—out to the dumpster with the contents of the litter box and a half bag of food. Before I started out the door, I placed a handful of her food into the pocket with the treats.

Even in death, Callie had one more lesson to share with me.

Our area in Phoenix is a mixture of desert landscape and grass to remind us of other climates. In front of our townhome is a wide, dry wash which is a mixture of cactus and grass.

Returning from the dumpster, I sat on the low retaining wall that overlooks the wash. I talked with her for a few minutes and then I took the food and treats from my pocket. "These are for you, kiddo. Whenever you get tired of playing in the grass, there will always be something here for you to eat." With a wide swing of my arm, I scattered the mixture out over the wall into the landscaped area.

Then I remembered the tuft of fur and removed it from my shirt pocket. "No one out here to chase you, Callie…only a field of grass to play in. You'll never run out of grasshoppers and crickets to chase." I blew the fur from my hand and it drifted toward the wash.

Callie was always a house cat, but I could almost see her chasing through the lush grass—ears laid back, her head out and a ruler straight line from the tip of her nose to the end of her tail. Now she was in a better place where her heart would not cause her pain, and she could run as long as she liked. I knew she would have no problem finding the food or the treats when she wanted to rest.

I said, "Goodbye, Callie" and walked slowly back home. The lesson she taught me was that saying "goodbye" is a process, not an event. She died here in our computer room just a few feet from where I am sitting now. It brings both tears and smiles as I write and remember.

The pain is fading, and more often than not I remember the dry wash and smile. We only lasted three days without a cat in the house. Saturday following Callie's death, we made a trip to the Humane Society's shelter. Shandy, a small, male, Maine Coon has taken over our home. He's another story, for another time.

"Callie, there's another 'critter' roaming your old domicile. Somehow, I don't think you mind because you know he is not a replacement. He's simply another in a long line of felines who will dominate the house as you did. I will miss you and I will always remember you. I feel the pain less often, and the smiles are more frequent. So, here's to tall grass, food, and treats. Thank you, and…goodbye Callie."

The End

Previous contest entry

AWARDS AND OTHER STUFF
by John Achor

Previously published, "Warrior Writers—Jouneys to Healing," by Humanities Nebraska (copyright retained).

I have a thousand memories and mementos from the military which mean a great deal to me. I won't include my wife and children; they are at the top of the list—a given.

I earned a dozen Air Medals, most of them for a set number of missions hanging around the coast of countries like Vietnam and Russia. Included in the group are also a few flights flown against the French nuclear tests in the South Pacific. One Air Medal was awarded for a single flight where we captured the signal of the HH radar, a suspected, yet previously unknown Cold War radar. According to the awards criteria, this action qualified for the Distinguished Flying Cross. However, politics being what they are—yes, even in the Air Force—the grand gurus of awards made this one an Air Medal. Though all the Air Medals appear the same, I'm proud to have piloted the plane when those guys in the back captured the new radar signal.

In the summer of 1969, I took one of the Alaskan RC-135Ds to Kadena Air Base, Okinawa — now part of Japan. My crew and plane were to fill a gap in the RC-135M fleet stationed there because those aircraft were long overdue for major overhauls. The M's had fan jet engines; we didn't. They could refuel in level flight after takeoff; we couldn't. These conditions caused headaches and heartburn for the staff, but as usual the crews get the job done; despite the staff. The missions were nineteen and a half hours long and we flew, without crew augmentation, every three days for the best part of a month. We completed our assignment, but this accomplishment was not the one affording me the fondest memory.

Back in those days, one of the popular half-hour TV sitcoms was "F-Troop." The show starred Forrest Tucker as a U. S. Calvary Sergeant in the old west. Our Kadena maintenance crew nicknamed themselves after the television show. I own a small, handmade certificate making me an honorary member of F-Troop—in my mind, it's a real tribute.

Leaving my Alaskan assignment, the Security Service Maintenance Commander presented me with a small award—a 5" x 7" framed photo of one of our planes. It was overlaid with a Security Service patch, a thank you and it was signed by the "Chief Wire Bender." Another memento that rates a spot on my wall of honor.

I received a similar award when I left the Joint Reconnaissance Center in the Pentagon. The standard presentation was a fancy 9 x 12-inch photograph of the Pentagon with space around the sides for others in the JRC to add comments and signatures. I appreciated the effort, but there was another that means more to me. Another 5" x 7" award, hand drawn and typed. It was signed by all the officer and NCO members of the Watch Team I worked with. There's a spot on my wall for this one as well.

Do self-awards count? In this instance, I think this one does. In the olden days, the steering wheels of cars were held in place with a single nut, and to cover it, builders installed a steering wheel cap. Be patient, all will reveal itself in due course. During my tour in Alaska, we spent about a week each month at Shemya Air Base in the Aleutian chain of islands, flying a one-of-a-kind airplane. The crew chief and maintenance staff believed she was the Cadillac of the fleet. They added a Cadillac steering wheel cap to each of the pilots' yokes. Landing at thirty minutes into the day of January 13, 1969, I found, later, the runway was too slick for us to stop. We went over a forty-five-foot cliff, but all eighteen on board walked away.

The crew chief entered the wreckage and salvaged those Cadillac wheel caps and gave them to me. I asked if they could procure a piece of the aircraft skin. They presented me with a chunk of aluminum about two feet by three feet, almost too big to stash in my A-3 bag—a rather voluminous zippered canvas container for flight gear.

In my basement back at Eielson Air Force Base, I cut eighteen three-inch squares from the aluminum skin. I fashioned and stained eighteen oak plaques. To every plaque, I attached an engraved plate containing one person's name and the date "13 Jan 1969" as well as a piece of the airplane skin. Two of the plaques were larger than the rest so they could also accommodate one of those Cadillac wheel caps. One of these went to my copilot and I kept the other. Another reserved spot on my wall.

Speaking of my copilot on that flight, after losing track of one another's whereabouts, we located each other on the internet a couple of years ago. I always knew this Air Force Academy graduate was a talented officer and if he stayed on active duty long enough, he would outrank me—he did. I always did my best to pass on knowledge and skill to those who worked for me, but never thought of it past that. In his email note to me, he included me in a group who affected his life and career—and considered me a mentor. Awards are not always tangible objects with a spot on the wall.

This final one needs a bit of background also. Before I moved to my Alaskan assignment, I attended an advanced Survival School at Fairchild Air Force Base at Spokane, Washington. This was a special course designed for strategic reconnaissance aircrews with much of the material classified above Top Secret. Even the names of the clearance levels were classified. To be a member of the school staff required clearances to these levels and extensive background checks. One of the senior Non-Commissioned Officers was a Master Sergeant—which as I recall, was the highest NCO rank available at the time. He began his military career during World War II in the German Wehrmacht. When his unit was overrun on the eastern front, he was conscripted into the Russian Army. The war over, he crawled under triple concertina barbed wire and made his way to U. S. occupied territory.

Proceeding to the United States, he joined our Air Force and rose to the rank of Master Sergeant. His distinctive voice, a growling guttural sound and his trademark gesture were ominous. From a fist, he extended his first and second fingers pointing forward and curved downward like a serpent's fangs. There was a sound to accompany the hand gesture and we called him the Snake—not to his face as I recall.

Our accommodations were sumptuous. An abandoned underground ammunition bunker was *home,* and was modified with "rooms" made of plywood. The cubicles extended the length of the bunker on both sides and a door which could be locked from the outside. Each compartment was tall enough to stand erect, but only thirty inches square. Resting was accomplished by leaning into a corner. We were not allowed to communicate with other "captives."

To be sure we remained silent, guards slipped into the pitch black quarters and listened. Offenders were removed from their cells and subjected to extra interrogations. On occasion, the Snake would enter and make his presence known. Even recognizing his voice, and knowing he was really on our side, a chill went up my spine at the sound of his voice.

Part of our training took place in a group compound. The course was too short to prepare a full-blown escape, but we were to make a concerted effort to plan one. On occasion, the guards would put canvass bags over our heads and march a line of hooded prisoners—who knows where. On other occasions, we were selected for interrogations. Placed in a small room with a table, two chairs and the inquisitor, we were to avoid giving up any substantive information to these interrogators.

At the end of the course, each student was debriefed by one of the staff. I drew the Snake. Rather than sitting behind the desk, he slid his chair to the side. The position allowed him to use the corner of the desk and sit closer to me. He covered all the items on the out-processing checklist and then we chatted for a few minutes. At last, he held out his hand to me. As we shook, he said, "I would work for you."

We stood and I left the room. I hope I at least muttered a "thank you," but I'm not sure. From a man with such a background, I consider his words as high praise indeed. It was not a tangible item for my wall, but I think it is the greatest award I ever received.

The End

Author's note:

I used Sepp's words; "I'd work for you" and inserted similar prose in my thriller, "Assault on the Presidency." Walking point—infantry jargon for the one designated point and leading the unit, is responsible for safety of the unit, looking for trip wires, ambushes, danger, etc. Hillbilly is describing his team and says, "I trust any of them walking point."

APPENDIX C – AUTHORS I MET OVER THE YEARS

Name dropper. Blog posting on August 4, 2015

Over the years, I've had the good fortune to meet a ton of well-known authors. I decided to "name drop" and put the list here. I met the vast majority of these authors at book signings at the Poisoned Pen Bookstore (an independent mystery book store) in Scottsdale, Arizona. Others were attending conferences or seminars.

After each name I added a bit of info about their writing, the name of a book current at the time or character names. Hang on, here we gooooo…

Lawrence Block (Matt Scudder, Bernie Rodenbarr, Vic Tanner)	James Lee Burke (Dave Robicheaux)
Laura Parker Castoro ("A New Lu")	Lee Child ("The Killing Floor")
Michael Connolly (Harry Bosch)	Clive Cussler ("Raise the Titanic")
Jeanne Dams (American writing British cozies)	Jeffery Deaver ("The Bone Collector")
Janet Evanovich (number series, "Seven Up")	Vince Flynn ("Memorial Day")
G. M. (Jerry) Ford ("Who the Hell is Wanda Fucha?" – Leo Watterman-4 homeless alcoholics)	Sue Grafton (alphabet mystery series, Kinsey Milhone)
L C Hayden (Aimee Brant and Harry Bronson mystery series)	Dame PD James UK author, Adam Dalglish
Laurie R. King ("Beekeepers Apprentice," Sherlock Holmes is the apprentice)	Steve Hamilton ("Ice Season")
Charlaine Harris (several series and TV "True Blood")	Tony Hillerman (SW, Navaho lawmen, Joe Leaphorn & Jim Chee)
Mary Lynn (mss prep and style, The Writer contributor)	Dennis Lehane ("A Drink Before the War")
Gayle Lynds ("The Last Spymaster")	Val McDermid ("Mermaids Singing," 2 other series)
Michael McGarrity (SW series, "Tularosa")	David Morrell (Rambo, "First Blood")
Sylvia Nobel (Kendall O'Dell mystery series)	Carol O'Connell ("Mallories Oracle")
Thomas Perry (Indian woman who helps people disappear, "Dance of the Dead")	

Pictures of several of these folks are on the next few pages.

First Thriller Conference, Phoenix, Arizona. After a couple of years, it was hijacked to NYC, by the Mystery Writers of America...we lived in Arkansas at the time...didn't make the NYC editions.

Pat is talking to Lee Child; author of the Jack Reacher books.

This is Vince Flynn, a thriller writer, who is autographing a copy of one of his books for Pat.

Vince died way before his time from cancer.

Clockwise from top left. John with Lawrence Block. Pat having a book autographed by Thomas Perry. Val McDermid, a British author who presented a seminar for a small group and was sponsored by Barbara Peters, owner of The Poisoned Pen Mystery Bookstore and The Poisoned Pen Press in Scottsdale, Arizona.

Val spoke about developing characters and she writes a male protagonist. Her presentation spurred my interest to develop my own character; I did and a year and a half later I breathed life into Acacia Fremont. Casey struggles to regain her dignity after a skunk of a husband has stolen her self-confidence. She takes temp jobs when the month is longer than the money, and finds herself in the middle of murder and mayhem.

Pat and G. M. Ford at The Poisoned Pen Mystery Bookstore.

He's really hamming it up.

Below is Pat with Jeffery Deaver at the first Killer Nashville conference. I caught him in the lobby, started a conversation and called Pat on a cell phone. She was in our room with two copies of his latest book; she rushed down to the lobby and he graciously signed both of them.

His picture on jacket covers make him look dreary and dower. As the keynote speaker we learned he has fantastic presentation and sense of humor. Don't judge the person by the book cover 😊

Michael Connelly is signing his new mystery for Pat. His Harry Bosche books were also made into TV series.

After the signing, he did a presentation for a small number of writers. He's the first I heard say, "To write, you have to know all the rules. Then you can break them."

This is Janet Evanovich, author of a numbered series of mysteries, featuring Stephanie Plum. She was signing her third best seller. If you've ever read her, you will be in love with Plum's Grandma Mazer, who attends funeral just for the fun of it.

In the top photo, Sue Grafton, the alphabet series author was signing books at The Poisoned Pen Mystery Bookstore. We had been in line for over two hours when we got to the table to get an autographed book. Even though she had another hour of signing ahead of her, she stopped to have this picture taken with her. Unfortunately, Sue passed away before she reached
"Z is for …"

Bottom, is the first writer's critique group I attended. We met every Monday at a local bookstore in Phoenix, Arizona. I learned more about writing from these folks that from anything else I did. Thank you, folks. Front row (l - r) are Judy, Wilma, and Val (facilitator), Back row
(l - r) are Cindy and John (that's me). A couple of members were missing that night. This is the group that presented me with the wooden "book" with built in clock, which always has a spot of prominence in our household.

APPENDIX D – ITEMS OF A MORE PERSONAL NATURE

NOTE TO SELF

John, you're in your mid-teens and you have a long life ahead of you. You are still a shy, introverted person; but that's not all bad. You have a lifetime to come out of your shell and you will—not that you'll become a roaring extrovert, life of the party person, but you will succeed.

You will find that you never can forecast what consequences small decisions will have, how they will affect you in later years. So, make those decisions as best you can, based on the best information you have at hand. Following that edict, will stand you in good stead.

By now, you've learned to be a loner not by choice, but by circumstance. In the first grade, you and a best friend survived being run down by a careless driver—with minimal injuries. Unfortunately, he came down with rheumatic fever, was isolated and that was the end of the friendship because you began your nomadic life style before he recovered. You will attend five grade schools before entering high school; another impediment to lasting relationships. You will have friends in high school and later, but never a chance of having those "lifelong" relationships so many speak of.

While it may be disappointing not having longtime friends, you will discover an advantage to being a nomad. In retirement years, you will meet people of advanced age who live today within fifty miles of their birthplace. You will also find these folks have an extremely narrow view of the world. Your grasp of a global understanding of economic and social effects on the population will be far beyond that of your neighbors.

You will enter college in June following high school graduation. The Korean era draft was in effect, and taking AFROTC in college will provide you a deferment until graduation. It did and you will receive your Bachelor's degree, be commissioned a Second Lieutenant in the United States Air Force, and best of all, be married all in the same week. You will enjoy six months of civilian life before being called to active duty and attend flight school. You will fly airplanes from the J-3 Cub to four-engine jets, from trainer to fighter to cargo to reconnaissance planes.

Your piloting skills will be challenged one night on a small island in the Aleutian Island chain. You will meet that challenge and be able to say, "All eighteen of us walked away."

In the Air Force, you will have many close associates, friends and crew members, however none will be in that super close friend category. Your three children will be born all across the country—Phoenix, Arizona, Falmouth, Massachusetts, and Spokane, Washington. Your nomadic career will allow you to live in or travel through all but one state in the United States—poor Vermont; your wife's birth state. These travels will take you to ten countries from Canada to Mexico, to England, to Japan, to South Korea, and Thailand; and to islands as small as Johnston atoll in the South Pacific to Greenland; to several oceans including the Pacific, Atlantic and Arctic; and over a dozen and a half dozen other major bodies of water—seas and gulfs, including the notorious Gulf of Tonkin.

You will be exposed to extremes of all types of life experiences and you will learn that in order to enjoy the peaks you have to endure the valleys. If life were to be a level, flat existence, it would be like the man who said: if I am careful enough, nothing good or bad will ever happen to me. If you never live through the depths of anguish, how will you ever enjoy soaring with the eagles? You will pass through both; and you will survive and grow.

One of the last chapters you will endure will be a world-wide pandemic. The end of this final slice of life has yet to be written. It will be a scary time, but I believe it will have a positive outcome.

Enjoy a long and prosperous life, you will have earned it.

LETTER TO FIVE GRANDDAUGHTERS

September 2, 2020

To my granddaughters: Bree, Emily, Katy, Lizzy, and Mirelly

You are living in tumultuous times. Seems like no matter where you look nothing will ever be *normal* again. But it will! It may be a different normal, but we all will get there.

It is hard to accept not being able to go where you want, be with who you want, do whatever you want to do. Remember, the restrictions will not last forever—it just seems that way.

Being deprived of everyday activities we are used to, hurts. It's not like sticking my finger hurt, but an inside hurt—one that makes me sad. And, it's okay to be sad, to shed a tear, as long as we remember the world will get better.

When I am missing something or someone, I like to think of Joe Biden speaking of the loss of his children. He said, "One day, thinking of them will bring a smile to your lips before it brings a tear to your eye." I call it *turning down the volume.* While the pain we feel will never go away altogether, nor should it, with time we learn to turn the volume down and then, perhaps that smile comes to us first.

And…one day all the distress and anxiety we feel will be in the past. And a day will come when we talk about this trouble and smile, even laugh. Today that may sound unbelievable, but I guarantee it will come true. So, today take care of yourself one day at a time. Then one day you will know that what I've told you today has come true.

As hard as today feels, there will come a day; while listening to *your* grandchildren who are complaining about the woes of their days, a memory will arise—a painful memory. And that smile will come to your face and a laugh to your throat. You will look at those grandchildren and say, "You think that's bad; you should have been there in 2020!"

When a time comes that everything looks its worst, feels like there is no answer, reach out, and no matter the distance, and take my hand. We will find the way to walk the path together. Enjoy yourself, be free and relish life to its fullest.

I love you, Grandpa Achor

JOURNAL TRAVEL

U.S. Forty-nine states; missing Vermont (Pat's birth state)

Foreign countries and possessions (* overflown)

Canada (5)
    Alberta, Jasper, Banff, Calgary, Lake Louise
    British Columbia, Vancouver
    Newfoundland, St. John
    Labrador, Goose Bay
    Ontario, Ottawa
Denmark*
England, Upper Heyford, Oxford, London
Greenland
Guam
Japan
Johnston Island
Mexico
    Ciudad Juarez
    Matamoros
    Nuevo Leon
    Reynosa
Okinawa (now Japan)
Philippines
South Korea
Sweden *
Taiwan
Thailand (14, 2*)

Bodies of water

Atlantic Ocean	Kara Sea
Arctic Ocean	Labrador Sea
Baltic Sea	Laptev Sea
Barents Sea	North Sea
Bering Sea	Norwegian Sea
Caribbean Sea	Pacific Ocean
Chukchi Sea	Philippine Sea
East China Sea	South China Sea
East Siberian Sea	South Pacific Ocean
Gulf of Alaska	21
Gulf of Mexico	
Gulf of Tonkin	

PLACES I/WE LIVED — one month to 14 years (# of times)

Alaska (1)	Arizona (3)	Arkansas (1)
California (2)	Florida (4)	Indiana (4)
Massachusetts (1)	Mississippi (2)	Nebraska (2)
Nevada (1)	Oklahoma (1)	Texas (5)
Virginia – Pentagon (1)	Washington (1)	South Korea (1)
Okinawa (1)	Thailand (1)	

My last assignment, from Jun 1973 to Jan 1976, was at The Pentagon where flight time was waived. I didn't need to fly to collect my flight pay. I accumulated a bit over 4,100 flying hours in my 18 years on active flight status. My time in the Arctic and South East Asia gave me over 900 hours of Combat Support time and 300 hours of Combat time.

SCRAPS OF PAPER

Over the years I've tried to journal; it didn't stick. I wound up with folders and notebooks full of scraps of ideas to include in this tome. I wasn't able to get to or find them all, and that's no doubt a positive circumstance—else this book would be three or four times longer.

So, this book became a mix of memory and stream of consciousness narrative. I know I've repeated bits and pieces because that stream sometimes runs over rapids that jog the memory in a different direction. As I edited, I decided to leave most of the repeats in since they came to me at a different time and under different circumstances.

Another of those asides: my work is reference intensive to make it easier to locate specific parts of my life by page number. Kindle Direct Publishing (KDP by Amazon) may impose pagination rules that throw my numbering off. I will do my best to keep it all straight, but if a reference is off a page or two, please bear with me.

I spent my life as a loner, an introvert—not by choice, but more by circumstance. My very early days were spent in isolation due to the location of our home and my best friend—six blocks away—coming down with rheumatic fever and being bedridden—and then we began lifetime nomads.

Again, relatively early in life, I came across the story of The Wandering Jew. He was consigned to an eternity of reincarnation always looking for something or somewhere. I always felt an affinity with that character. Within the past year or so, our son Mark had one of those gene/DNA tests run to tell where your ancestors came from. Guess what? Our ancestors seem to be of Central European Jewish origin. The name Achor, appears in the KJV of the Bible twice: book of Joshua where the valley is described as "muddy, turbid: gloomy, dejected." But on a positive note, is added, "God made the place of trouble into a door of hope." In 1 Timothy, the reference is more obscure, but I got the feeling that "Achor means trouble."

Being a spiritualist, not religious, I'll take the references with a grain of salt. Especially since the spelling of "Achor" seems to vary here and there.

I inherited my loner status from my father. He lost his father at sixteen and had minimal male influence. I remember at a party, I looked around for Dad and found him in solitude in a darkened room looking out a window at the night sky, and thinking—who knows what? Like most of us, he did the best with what he had; he just didn't have the background to do a better job. I think in many ways, I suffered the same fate.

I remember an episode of the TV show, "Blue Bloods," a cop family, grandpa was and dad is the Police Commissioner, three male children are cops and a daughter is a DA. One son was killed in the line of duty. In this particular show, the youngest son had a problem and goes to grandpa for counsel. He receives sage words of wisdom, and it struck me that of four grandparents, I only knew one—my maternal grandmother.

I suppose I need to close with a few words about my last career, being an author. If Pat and I had needed my writing income to exist, we both would have starved to death years ago. I wish I had been as successful at some of those folks I showed you in this Appendix.

Getting published is a grand turkey shoot. Timing and circumstances are against all but a few. Regardless, I enjoyed the time I spent writing characters I dredged up out of my mind and the hours pouring out their tales into the computer.

My male protagonist, Alex Hilliard, was fairly easy. He was male, a USAF reconnaissance pilot, was stationed in Alaska and the Pentagon, Sound familiar. With all the similarities, he was far more adventurous than I.

I mentioned that Val McDermid's words encouraged me to develop my female protagonist, Acacia (Casey) Fremont. If we had not moved, her adventures would have occurred in Arizona. As it was, it was easier to place her in Little Rock, Arkansas. Even though a self-styled "expert" lady told me it was impossible for a man to write a woman's character. Stephen King could give her a good argument. I think I did a creditable job writing Casey, and only one time in a critique group did a lady hear a few of my words from my protagonist, and said to me, "John, that's a guy thing." 😊

There are a few more problems cropping up that I covered in the Note to Self, but all in all, it's been a grand ride. I married the greatest gal in the world, who gave us three fantastic children who I'm proud to be their father. I hope I provided them with a modicum of wisdom along the way. I did the best I could with what I had; much like we all do.

I hope you found some of my life of interest as you plowed through nearly 300 pages that cover almost 90 years of my life. I probably enjoyed it more that you did.

Here is a last scrap of paper I did find. It's the story of, Rhysling a blind space traveler scrounging any seat available on any spacecraft that could get him home. My best guess is that I heard the story on a Saturday morning Sci-Fi radio show. I just located the source; it's from a short story, "The Green Hills of Earth," by Robert Heinlein and was written in 1947. With an attribution to the author, I'll repeat the words I heard almost four score years ago?

> I pray for one last landing,
>
> On the planet that gave me birth,
>
> To rest my eyes on the fleecy skies,
>
> And the cool, green hills of earth.

So, whether I am The Wandering Jew or Rhysling, my journey is drawing to a close. It has been a hell of a ride, because I was fortunate to be surrounded by a fabulous family, great friends, and all the people I've met along this journey. It took me nine weeks to fight off and rehab from COVID-19 and then I handled knee surgery and the appearance of prostate cancer. So, I still have a couple of battles going on but I'm looking for positive outcomes.

Goodbye, to my friends and readers, I love you all. I can only hope that you found a smile and a tear along the way and enjoyed, to some degree, this manuscript that I've poured out on paper.

It's tough for a writer to find that last "Three-Oh mark" for a story, because we love the act of putting words on paper. It is hard, but this is the time for that final scrap of paper and the end of story sign off…

John

= 30 =

Made in the USA
Columbia, SC
10 February 2023

11800454R00173